Naturalists' Handbooks 31

Amphibians and reptiles

TREVOR J.C. BEEBEE

Pelagic Publishing
www.pelagicpublishing.com

Published by **Pelagic Publishing**
www.pelagicpublishing.com
PO Box 725, Exeter, EX1 9QU

Amphibians and reptiles
Naturalists' Handbooks 31

Series editors
S. A. Corbet and R. H. L. Disney

ISBN 978-1-907807-45-9 (paperback)
ISBN 978-1-907807-46-6 (ePub)
ISBN 978-1-907807-47-3 (mobi)
ISBN 978-1-907807-48-0 (PDF)

Text © Pelagic Publishing 2013

British Library Cataloguing in Publication Data
A catalogue record for this book is available from
the British Library.

Printed and bound in the UK by CPI Colour.

Cover photographs: great crested newt (Photoshot
Holdings/Alamy), grass snake (Fred Holmes); common
frog (Erik Paterson, www.erikpaterson.co.uk)

Contents

Foreword

The uncoiling rush and sliding olive of a grass snake, the still, unblinking cinnamon eye of a silver grey adder and the emerald flash of a darting sand lizard through a cushion of heather are amongst the most tantalising and thrilling glimpses of our native wildlife you can find. The slow, deliberate posturing of a dragonised male smooth newt fanning his female, the perverse fervour of spawning common toads as they roll in cold ecstasy on a spring night or that first encounter on a wet sandy slack with the call of the natterjack are all encounters to be treasured. And yet our reptile and amphibian fauna are sadly too often neglected by naturalists. They are happened across rather than sought and sidelined behind birds, butterflies and the likes of orchids. This is such a shame, as they are both beautiful and fascinating animals.

I entertained a huge crush on these groups as a child, indeed I still count grass snakes as one of my favourite animals. Peering into murky or dark ponds, sneaking through heather and gorse looking for their resting spots, or twisting quickly to hear the characteristic rustle of a fleeing snake or lizard was all of my life between about eight and ten. I still have the first sloughed adders skin that I delicately teased out of some brambles all those years ago.

This book, by one of the UK's foremost authorities, seeks to make this small guild of underestimated creatures more accessible through simply highlighting the ease of identifying our reptiles and amphibians and fuelling a greater interest in their behaviour and ecology. They are not enigmas, they are actually quite easy to find and with a modicum of practice equally easy to watch and enjoy. I seriously hope you take up the challenge because, unfortunately, a number of these species are either rare or in decline and it is only through engagement, actually meeting an animal, that any lasting affinity for it can be forged. This passion is essential to motivate and maintain our conservation efforts. So, charge your torches, get your wellies out of the cupboard, don't forget your binoculars and camera, and most important of all, pack your children into the car. Then get out, and get to grips with our fabulous snakes, lizards, frogs, toads and newts.

Chris Packham

Acknowledgements

Thanks to Tony Gent of Amphibian & Reptile Conservation for suggesting this book, to Nigel Massen and Sally Corbet of Pelagic Publishing for advice and support, and to Ronn Altig, Brian Banks, Lee Brady, Tim Harry (Chaffinworks), Richard Newton, Charles Snell, Paul Wells, Julian Whitehurst, John Wilkinson, Andrew Williams (Critterzone) and Wolfgang Wuster as well as contributors to ARC's photo archive (Neal Armour-Chelu, Chris Dresh, Tony Gent, Chris Gleed-Owen, Fred Holmes, Howard Inns, Rodger McPhail, Nick Moulton and Angela Reynolds) for supplying photographs. Additional images were supplied by: Erik Paterson (www.erikpaterson.co.uk/), Kelly Minars (Flickr/greaterumbrage), Keri Leaman (Flickr/birderkeri), Quentin Scouflaire (Flickr/sqfp.info), Duncan Hull (Flickr/dullhunk), Alan Martin and Steve Ogden (www.wildlifeinsight.com).

Author royalties from this book have been donated to Amphibian and Reptile Conservation.

1 Introduction

1.1 Interest in amphibians and reptiles

Amphibians and reptiles have a long history of both fascination and, sadly, repulsion in the common imagination. Although often considered together by naturalists and the public alike, in fact these two groups of animals have little in common apart from being vertebrates. Perhaps the simple fact that none of them are warm and furry (mammals), feathery fliers (birds) or streamlined swimmers (fishes) is sufficient to explain why most people think of them collectively. In the not too distant past even professional zoologists made the same mistake. Until the early nineteenth century, for example, lizards and newts were not properly distinguished. We still have the scientific discipline of herpetology which continues to pool amphibians and reptiles together, and herpetological societies around the world that do the same thing. The number of enthusiasts for these animals has remained relatively small compared with the other vertebrate groups and so far there simply aren't enough to warrant the separation that taxonomic distinction justifies. This book continues the tradition of treating amphibians and reptiles together, partly for an additional reason. In Britain there are, even combined, far fewer species than in any of the other vertebrate groups. Fortunately what they lack in species diversity the British amphibians and reptiles more than make up for in intrinsic interest and accessibility for study.

In general amphibians and reptiles were largely ignored by early naturalists, including the pioneering taxonomist Carl Linnaeus who declared a profound dislike of them. Even Gilbert White gave scant attention to frogs, newts, lizards or snakes in letters describing wildlife around his Selborne parish in the eighteenth century, an area that was and still remains home to all the British species. These people didn't know what they were missing. Happily things have changed dramatically over the past fifty years or so and there is now a much greater interest in these fascinating vertebrates, to the extent that since 1989 there have been regular world congresses of herpetology attended by hundreds of scientists from every corner of the planet. What has triggered this change of fortune?

The answer to that is a mix of intriguing recent revelations. Every schoolchild learns about frogspawn, tadpoles and the dramatic events of metamorphosis resulting in tiny frogs and toads around the pond edge. But it has become clear that this only represents one method of reproduction, albeit the commonest type in temperate countries. In the tropics things are very different. Some frogs lay eggs singly in the small volumes of water held among the leaves of bromeliads, plants that grow on tree branches high in the forest canopy. These frogs return later to feed their tadpoles with un-fertilised eggs. Others rear tadpoles in their mouths, on their backs or even embedded within their skin. Females of one Australian frog, now extinct, swallowed their eggs and allowed the tadpoles to grow in their stomachs before regurgitating fully-formed froglets. It turns out that frogs have a greater variety of reproductive methods than any other animal group. And it's not just the basic biology that has attracted attention. Darwin recognised the importance of sexual selection in evolution and frogs demonstrate this in quite extraordinary ways. In many species the strength or frequency of male croaks makes all the difference to an individual's chances of attracting a female and therefore of successful reproduc-tion. In newts there is a comparable situation: the large and striking crests of males in the breeding season are used to impress potential mates in elaborate underwater displays (Duellman and Trueb, 1994*).

Sexual selection

Sometimes regarded as a form of natural selection, sexual selection is an evolutionary process that depends on competition among (usually) males for access to females, or on preferences among (usually) females for particular male traits

*References cited in the text appear in full on page **165**.

Reptiles are equally remarkable in their breeding behaviour. Male sand lizards turn a beautiful iridescent green in spring and compete for females by a combination of aggressive displays and physical fights. Amazingly, the intensity of the green colour can be temporarily enhanced during these encounters to increase intimi-dation of a rival, or maybe to impress nearby females. These lizards have provided some of the best examples of the benefits of multiple paternity in nature. Females usually mate with several males and those that do this tend to have the fittest offspring. For all British species of amphibians and reptiles, spring is the season when they are most visible because this is when breeding occurs with its associated extrovert behaviours. Each species has its own special effects. Male adders 'dance' together in power struggles, preoccupied with winning the contest and utterly indifferent to anybody watching them. 'Balls' of grass snakes thrashing around in the herbage are made up of several males wrestling for access to a (usually) much larger female somewhere in the middle. And so on. Vitt and Caldwell (2009) summarise much of the knowledge and excitement concerning both amphibian and reptile biology.

It's not just their sex lives that have increased interest in amphibians and reptiles. These animals play pivotal roles in food chains, both as predators and as prey. All are ectotherms with no need to use energy for maintaining body temperatures, so they can get by with very little food. One consequence of this is that many species exist at much higher densities than comparably small mammals like mice, voles and shrews. The sheer numbers of toads emerging after rain on a warm summer evening give an idea of how important they must be both to their main prey (various invertebrates) and to animals that prey on them, of which there are a lot. Abundant and diverse amphibian and reptile faunas are good indicators of habitat quality and biodiversity in general and have attracted the attention of ecologists for that very reason.

Ectotherm
An animal dependent on environmental temperature and unable to generate its own body heat

All of which leads on to a related point, the realisation that amphibians and reptiles have declined around the world over recent decades more dramatically than any other vertebrates (Stuart and others, 2004; Beebee, Wilkinson and Buckley, 2009). We obviously need to know why this is happening and try to minimise the damage. Whatever the causes are, there is no reason to suppose that other groups, including mammals like ourselves, will be perpetually immune to them. Arguably this is rather a gloomy reason to be interested in these animals but the importance of the subject is beyond dispute, and study of declines can provide surprising insights in unexpected ways, such as the impact of new 'emerging' diseases on wildlife populations. These studies can also have early practical consequences, for example by showing conservation organisations how best to improve a habitat. Sometimes such management, designed for a particular species, has unexpected extra benefits for others. Pond restoration for natterjack toads on a Hampshire heath, for example, revived populations of shoreweed (*Littorella uniflora*, a rare plant) and the spangled water beetle (*Graphoderus zonatus*) at its only British locality.

Finally it's worth mentioning that amphibians and reptiles can sometimes provide direct benefits to humans, another source of interest in these days of 'ecosystem

Ecosystem services
Features of the natural world that contribute to human wellbeing, such as pollination, or the uptake of nutrients by reedbeds

Viviparous
Bearing live young (so not laying eggs)

Fig. 1.1 Adult female common toad *Bufo bufo* (Erik Paterson)

services'. The toxins in toad skins have been investigated for their antibiotic properties (so far without much success) and crocodiles are farmed for their leathery skin and its contribution to the handbag trade. None of which, of course, is of much concern to most naturalists.

1.2 Engaging with amphibians and reptiles

The first questions you are bound to ask if you are thinking of studying these creatures include: where do they live and how can I find them? The good news is that few places in the British Isles are distant from at least one or two species of amphibians or reptiles. It will not be necessary to travel very far to locate some of the commoner species, notably smooth, palmate and great crested newts, common frogs and toads (**Fig. 1.1**), viviparous lizards, slow-worms, adders and grass snakes. All of these animals are habitat generalists and can be found, with luck, in a broad range of landscapes including farmland (though not so often where it is under intensive management), open woodland, heaths, commons, moors, sand dunes, parks and gardens. Of course there's more to it than just turning up somewhere and hoping for the best, and not all the common species occur in all these habitats. The rare native species (natterjack toad, pool frog, sand lizard and smooth snake) are by definition not so easy to find. Native pool frogs have a very restricted distribution, currently at just one site in Norfolk. The other three rarities are heathland and sand dune

specialists with very limited ranges, mostly in Surrey, Hampshire and Dorset for the reptiles but additionally at sites in East Anglia and around the Irish Sea coast in the case of natterjacks. Several additional, non-native species of amphibians and reptiles have been introduced into Britain over the past two hundred years but only three have become at all widespread. You might find noisy marsh frogs in low-lying ditch systems in various parts of southern and eastern England, especially in Kent and Sussex. Brilliantly coloured alpine newts now exist as many small colonies, mainly in parks and gardens, over much of Britain. And wall lizards thrive in an increasing number of favoured sites, such as south-facing cliffs and walls, mostly in southern England.

Looking for amphibians is very different from searching out reptiles, with one exception (described below). Common frogs, toads and newts spend most of their lives hiding in vegetation, in burrows or under stones and logs well away from water. This is where they hunt their prey, invariably at night, throughout the summer and autumn months. In winter they become even more secluded, usually in frost-free refuges well below ground. To find amphibians in these phases of their life cycle, the only options are to turn over likely hiding places (any sort of debris, natural or man-made, preferably on at least slightly damp soil); or to search at night with a powerful torch. This only works well if the habitat is fairly open, such as grazed pasture, heath or mobile dunes.

Neither of these approaches guarantees success, but all species are relatively easy to find when they are mating and spawning. For this reason most people engage with amphibians during the spring months when they resort to water for breeding. However, choosing the best ponds to search requires some knowledge of what amphibians are looking for. Large ponds with fish are often selected by toads but are unlikely to be good for any of the other species. Great crested newts in particular usually avoid fish ponds because their larvae are very vulnerable to fish predation. Smallish, fish-free ponds close to good quality terrestrial habitat (so not set in the middle of intensive arable farmland) are the best bet for all the other common species. For natterjacks, very shallow, temporary ponds on heaths and dunes are the favoured spawning sites.

Timing is also critical. Common frogs and toads are 'explosive' breeders and only visit the pond for a week or two in early spring, with toads usually arriving a little later than frogs. The precise timing varies a lot across Britain. In parts of Cornwall frogs regularly spawn before the end of December (hardly spring!) while in the highest Scottish mountains they may wait until early April. Newts are less problematic because they stay in the ponds much longer, for at least several weeks, often starting in February or March and remaining until May or early June. Natterjacks behave like newts in this respect. They have a protracted spawning season, usually concentrated in April and May, so much later than common frogs and toads.

The above-mentioned exceptions to the generally elusive nature of amphibians are the introduced marsh frog and its close (but rarer in the UK) relatives, pool and edible frogs, collectively referred to as 'water frogs'. These amphibians got their general name because they stay in or close to water all year round and can be seen sunning themselves, just like reptiles, on the banks of ponds or ditches from spring through to autumn.

Tadpoles offer further opportunities for engaging with amphibians. Those of frogs and toads are often easy to see in ponds and are present from spring to early summer, or late summer in the case of water frogs. Newt tadpoles are less obvious because they hide in pondweed but are also present for a long time, typically from late May until August or September. Sometimes they even overwinter in the breeding ponds.

Fig. 1.2 Typical reptile habitat (Tony Gent)

All British reptiles are active in daytime throughout most of the year, apart from the winter months when they hibernate underground. However, unlike amphibians they do not congregate in particular places (like ponds) for reproduction. This makes them generally harder to find. Spring is again the best period because this is when all our reptile species emerge from hibernation. They mate soon afterwards, and spend a lot of time basking in the early spring sunshine, and then engage in elaborate courtship rituals in which males in particular are bolder and less watchful for predators than they become later in the year. With experience it is possible to predict likely places for spotting reptiles, mostly when they are basking to warm up. Weather conditions are critical. In spring almost any time of day is suitable, provided there is some sunshine and preferably little wind. As the weather warms up, though, basking only occurs very early or late in the day. As with amphibians, refugia may also be useful, especially large flat items such as pieces of corrugated iron or roofing felt. Snakes and slow-worms often hide under such refugia, but viviparous and sand lizards rarely do so, and those species are best found when they are warming up in the sun.

Refugia
Hiding places

The best places to look for most reptiles are south-facing banks with low-growing vegetation such as heathers or short grass, together with some scrub (**Fig. 1.2**). Basking animals seek out small patches of open ground surrounded by vegetation into which they can escape quickly when disturbed or threatened. However, grass snakes may also profitably be sought in and around ponds and ditches where they hunt their amphibian prey. In areas where grass snakes are known to live, compost heaps can be good places to look if the timing is right. Females congregate to lay eggs there in early summer, and later on the newly-hatched snakes emerge, sometimes in large numbers.

1.3 Special aids

A couple of relatively recent developments have proved especially useful in the study of amphibians and reptiles for both amateur and professional investigators. The first of these has been the advent of high quality digital photography. With reasonably cheap cameras that come with inbuilt zoom and macro lenses it is now possible to take superb, detailed photographs of many types of wildlife. Amphibians and reptiles lend themselves very well to such photography because most species can be approached easily in the wild or caught and handled for close-ups. Just making a picture collection of the various species, including different behaviours and life stages, is rewarding in itself.

This is not, however, the main reason why digital photography has become so valuable in amphibian and reptile research. For many purposes it is useful to identify individual animals if they are seen or captured more than once. This kind of information allows estimation of home ranges and survival over months or even years. In the past this could only be achieved by marking the amphibian or reptile in some way, often with tags wrapped round part of the body or even by clipping off the ends of toes in an individual-specific pattern. This level of interference is undesirable for many reasons, not least of which is the possible risk of increasing mortality (for example from infection where toes are clipped). Passive integrated transponders, PIT tags, have also been employed with amphibians and reptiles. These are injected under the skin and each has a specific code that can be read by holding a recording device next to the animal. Although it is a substantial advance on earlier techniques, PIT tagging also has disadvantages. Usually the amphibian or reptile has to be caught again because the recorder only functions close up; PIT tags are expensive and too big to inject into young amphibians and reptiles; and there is still some risk of infection, although they come in sterile packages.

Digital photography has revolutionised the tracking of individuals because many of our amphibians and reptiles have unique markings of some kind that can be caught on camera. The spot patterns on the backs of lizards, scale arrangement and colouration around the heads of snakes, belly spots of crested newts and wart or back stripe patterns of natterjack toads have all provided 'personal' photographic fingerprints. This can even work for very young animals such as newly-metamorphosed natterjack toads, although patterns do sometimes change a bit as animals grow, requiring careful checks if identification is to be reliable over long periods. Another advantage is that photography is minimally invasive and sometimes individuals can be recognised by taking pictures in the wild (for example of basking lizards) without having to catch or handle the animal at all. It's not perfect – some species such as smooth and palmate newts have very few features that might be used for individual identification. But there's room for much more study along these lines, as will be discussed in later chapters. Software for automated screening of photographs is also developing rapidly, taking away the tedium of having to search large picture libraries by eye to identify recaptured individuals.

The second recent development, of special relevance to amphibians, is the vogue for installing garden ponds, which have become increasingly popular since the 1960s. Mostly these ponds are made for ornamental reasons and usually they are stocked with fish, but this has not prevented them from becoming important breeding sites for some of the commoner amphibian species. This may be because most garden pond fish are goldfish or other varieties of carp, and these are omnivores rather than dedicated predators, thus allowing some amphibian larvae to survive. Common frogs have benefited the most from garden wetlands, followed by smooth (and possibly palmate) newts. Common toads crop up less often and usually only in the largest ponds. Great crested newts are rare; even carp may be too much for them, and in any case they also prefer quite big ponds. Gardens are also key sites for at least two of the introduced species. Alpine newts occur almost exclusively in garden ponds,

Omnivore
An animal that eats both plant and animal material

Fig. 1.3 A garden pond used for breeding by several species of amphibians

and so do the much rarer midwife toads, which survive in just a few towns in England as far as we know. By the late 1970s about one garden in seven in the Brighton area had a pond and about half of these were used by at least one species of amphibian. This amounted to several thousand breeding sites for common frogs, smooth newts and common toads (Beebee, 1979). A similar picture has emerged in other towns and cities in Britain wherever anyone has bothered to look.

What all this means is that equipped only with a garden pond (**Fig. 1.3**) it is possible to study the breeding behaviour and larval development of amphibians within a few metres of your back door. Such convenience makes intensive study remarkably easy and there is scope for much more work using this resource. If there are local populations a new garden pond will probably be colonised within a couple of years, but there's no need to wait that long if other ponds with frogs or newts are available. Transferring small amounts of spawn (**Fig. 1.4**) or tadpoles (frogs) or a few adults (newts) can initiate a new population within weeks of finishing the pond construction.

Unfortunately gardens, even those managed with wildlife in mind, are generally less useful for studying reptiles. This is probably because there is usually too much disturbance for species that need to bask in daytime (that is, for most of them) and there are high concentrations of effective predators. Cats and blackbirds are remarkably adept at catching lizards. Grass snakes sometimes visit garden ponds but by far the most successful garden reptile is the slow-worm. These lizards sometimes occur in large numbers and survive because they are so secretive (though cats still catch a few), but this also makes them hard to study. Nevertheless there is the prospect of carrying out interesting work on this species in gardens, perhaps more effectively than in the wider countryside, simply because it's possible to work so intensively close to home.

1.4 Where next?

The primary focus of this book is to demonstrate how easy it is to carry out original work on amphibians and reptiles even without professional training and equipment. Despite the increased interest in both groups there is still a lot left to discover and much can be done with minimal apparatus. After some background material on the biology, ecology and conservation of amphibians and reptiles (chapters 2 and 3), the book introduces some topics on which further research is needed, and describes the hows, whys and wherefores of carrying out new studies on these intriguing and very amenable subjects. Of particular importance is the contribution that amateurs can make to the survey and monitoring of amphibian and reptile populations across Britain, a national exercise devoted to finding out how the status of each species is bearing up under modern pressures on the countryside (chapter 4). Chapters 5 (amphibians) and 6 (reptiles) outline various types of study that could make substantial contributions to knowledge about all our species, and chapter 7 proposes collaborative studies, in which schools could make a particularly valuable contribution. Chapter 8 consists of keys for identification, and is followed by some essential information including the requirement for licences for handling the animals, data analysis and publishing results in scientific literature (chapter 9), and useful addresses and links (chapter 10). Finally there is a reference list of relevant work cited in this volume. Other books that cover the biology and identification of British amphibians and reptiles in detail include a *New Naturalist* volume (Beebee and Griffiths, 2000), a well illustrated recent guide (Inns, 2009) and a publication dedicated to crested newts (Jehle, Thiesmeier and Foster, 2011).

Fig. 1.4 Frog spawn (Chris Dresh)

2 Basic biology

2.1 Life histories

It's in their life histories that amphibians and reptiles differ most strikingly. All amphibians need water for reproduction. In the tropics this requirement can sometimes be accommodated without resort to ponds or streams (usually by adults carrying tadpoles on or in their bodies), but in Britain access to a wetland is essential for all our species. Unshelled eggs (spawn) are deposited in water, without which they desiccate and die very quickly. The eggs may be laid singly and wrapped in the leaves of pondweeds, as in the newts, or laid in clumps or strings as in frogs and toads. Depending on temperature, development of embryos is complete within one to two weeks and the larvae (tadpoles) become free-swimming soon after hatch. An astonishing feature of great crested newt eggs is that 50% inevitably die before hatch, the result of a genetic anomaly not seen in any other British species. Larval development continues in the pond for as little as four weeks for natterjack toads in warm springs but occasionally for up to a year in the case of newts. This aquatic phase of amphibian life history is too well known to merit detailed description here but remains one of their most fascinating features, highly accessible to study and still open to new discoveries, as indicated in later chapters. Sooner or later, usually around midsummer for frogs and toads but in August or September for newts, the larvae metamorphose into miniature versions of the adults, and these juveniles immediately leave the water. The final hormone-triggered transformation is rapid and normally accomplished within a few days.

Anuran (frog and toad) tadpoles are omnivorous, consuming microorganisms in the water column but also grazing sediments on the pond bottom or even from beneath the surface film. Urodele (newt) larvae are entirely carnivorous and feed on whatever they can catch that is small enough to eat, especially crustaceans such as Daphnia. After leaving their natal ponds young newts carry on feeding on small invertebrates, whereas frogs and toads make a dramatic transition from omnivore to carnivore, now also pursuing invertebrate prey. For around two years, sometimes longer, juvenile amphibians of all species remain almost entirely terrestrial while they grow to adult size. Great crested newts are an exception because young animals occasionally return to ponds for a while in spring, presumably because these are a good source of food. In summer and autumn juveniles and adults alike venture out of their hiding places on warm, damp nights to hunt the small animals that provide all their food. In winter they seek shelter from frosts, usually underground but occasionally (especially in the case of frogs) at the bottom of ponds. Some time during the spring adults of all species migrate to their breeding ponds for spawning while juveniles remain on land.

Adult frogs, toads and newts can survive in captivity for many years, much longer than mammals or most songbirds of comparable size. Ages of over twenty have been recorded for some species, including common toads and great crested newts. In the wild, lifespans are typically much shorter. Although there is considerable variation among species and locations, a seven or eight year old amphibian is at the top end of normal life expectancy and the probability of surviving from one year to the next is commonly around 50%. Individuals in high altitude populations tend to live longer than those in the lowlands, possibly because they are active for fewer months each year and the hibernation period (when they are not exposed to predators) is relatively long. All this means that many individuals have more than one chance of reproduction and amphibian populations normally consist of overlapping generations.

The life history of a reptile is very different. Reproduction does not involve water because eggs, where present, are shelled and partly resistant to desiccation. They are not however as impermeable as bird eggs, being leathery rather than calcified on the outside. This means that egg-laying requires conditions with at least some moisture if the embryos are to survive. But even within the small British list there are distinct differences in reproduction mode. We have only two native species that are oviparous, notably the sand lizard and the grass snake. The lizard uses holes dug in sand whereas the snake selects warm spots, ideally with heat supplemented by fermentation of decaying vegetable matter, such as piles of dead leaves or compost heaps. Eggs develop through the summer and hatch in August or September. Our other species are viviparous, giving

Calcified
Hardened by deposition of calcium salts, as in bones

Oviparous
Laying eggs

birth to fully formed live young after the eggs have completed development within the mother. This strategy is adopted by viviparous (common) lizards, slow-worms, adders and smooth snakes. Often the juveniles emerge still encased in a membrane but wriggle free from it almost immediately. Viviparity in reptiles means that the mother can, by basking, maintain much better control over development than is possible where eggs are left to take their chances. On the other hand, the extra basking needed in summer may expose females to elevated risk of predation. Viviparity becomes increasingly frequent among reptile species at high latitudes where temperature may be most critical to reproductive success. It was surprising to discover that in some reptiles, especially turtles, incubation temperature influences sex determination of the offspring, but this has not been demonstrated in any British species.

Juvenile snakes and lizards disperse and, like amphibians, spend at least the next two or three years growing to adulthood. All our reptiles are predatory throughout their lives, though lizards occasionally lick flowers (presumably for nectar) and in captivity may relish sugar or honey. Generally, though, lizards prey predominantly on invertebrates and snakes mainly on vertebrates, including other reptiles, amphibians, small mammals and nestling birds. In spring adult reptiles mate but otherwise all age classes spend the warm months basking to warm up, hunting prey almost always in daytime and secreting themselves in refugia overnight and on hot summer days. The diurnal activity patterns of most retiles are therefore the complete opposite of those of most amphibians. In winter all our reptiles seek frost-free refugia for hibernation underground.

Reptiles resemble amphibians in that they are surprisingly long-lived for their size, but they show much greater interspecific variation. Slow-worms and smooth snakes fare the best, sometimes living for decades even in the wild. It may be no coincidence that these are the species that spend the least time above ground, even in summer. Both are secretive, remaining for long periods buried beneath soil or sand, and warm up mostly under at least partial cover. Viviparous and sand lizards rarely survive to be older than six or seven. Grass snakes and adders are intermediate, with some individuals making it into a second decade. Just like amphibians, reptile populations therefore also comprise overlapping generations. For both groups relatively good survival prospects from year to year mean that the occasional season with total breeding failure is not a catastrophic disaster.

2.2 Skin

Probably the second most widely cited difference between amphibians and reptiles is in the nature of the skin. In amphibians this is naked, liberally supplied with pores that generate secretions, and more or less freely permeable to water. This means that amphibians away from ponds or streams are always at risk of desiccation. There have been several evolutionary developments to minimise this danger. Species in extreme habitats such as deserts spend much of their lives encased below ground in water-filled cocoons waiting for rain; and in high forest canopies there are tree frogs which produce waxy secretions that minimise water loss. In Britain there is nothing so dramatic but nevertheless there are significant differences between species. Toads can survive in drier habitats than frogs, essentially because their skins are thicker. Amphibians do not drink but capitalise on their permeable skin by absorbing water through it, over the entire body, just by sitting in ponds. Natterjacks cope with the arid conditions of heaths and dunes by digging burrows down to damp sand. They have a large 'drink patch' of skin on the rear of their underside which is particularly efficient in taking up water from the substrate in these circumstances.

Amphibian skin (**Fig. 2.1**) frequently has an additional protective function. Without fur, feathers or scales the animals appear highly vulnerable to attack. To compensate for this, glands such as the conspicuous parotoids behind the eyes of toads secrete a wide range of toxic compounds. Species vary enormously in this regard. Common frogs, smooth and palmate newts have few toxins and suffer from a wide range of predators as a result. Frogs rely mostly on hop speed to escape. Water frogs don't have parotoid glands but still produce poisons in the skin, perhaps because being more diurnal than common frogs they are at extra risk owing to their conspicuous basking habits. Even their tadpoles have chemical defences, which are effective against small fish such as sticklebacks. Toads and great crested newts are the best protected of the British species, capable of producing copious secretions from glands and the warts covering most of their bodies. The bright 'wasp' colours

Fig. 2.1 Common toad showing parotoid glands (Erik Paterson)

Fig. 2.2 The warning colouration on a great crested newt's belly (Erik Paterson)

Fig. 2.3 Reptile scales (Kelley Minars)

of crested newt bellies (**Fig. 2.2**) are probably a warning of this to would-be predators. Toad tadpoles, like those of water frogs, gain some chemical protection from birth, though, curiously, great crested newt larvae do not. In any case, toxins are not an insurmountable defence. Some predators, including grass snakes, consume them with apparent indifference while others such as crows have learnt to discard the skin and extract the body contents. Toad tadpoles are unpalatable to most amphibian and fish predators but still fall foul of invertebrates like dragonfly nymphs which, again, essentially discard the toxic body parts.

By contrast, reptiles are protected with scales (**Fig. 2.3**) and in the case of the slow-worm, also with bony plates (osteoderms) like a suit of armour immediately beneath the skin. This tough exterior renders reptiles very resistant to water loss, allowing them to thrive in the driest of habitats. Unlike amphibians, reptiles drink through the mouth but many species get most of their water from their prey. They also produce very dry excreta, rich in uric acid, and thus minimise water waste by avoiding a need to urinate. Even so, it's not hard to find reptiles actively drinking from ponds or puddles when the opportunity arises. No British reptile produces skin toxins but grass snakes and slow-worms readily expel foul-smelling excreta from the cloaca when caught and handled, presumably as an adaptation to deter predators.

All amphibians and reptiles shed their outer skins (**Fig. 2.4**) periodically throughout life, normally several times each year. Frogs and toads peel the old skin off and push it directly into their mouth. Newts often do the same but sometimes just leave it, and newt 'ghosts' can be found floating in ponds during the breeding season. Lizards shed their skin piecemeal, in flakes that adorn the vegetation they move through. Snakes and sometimes slow-worms frequently shed their skin in one piece, turned inside out after it splits around the mouth, and peeled off backwards like a sock as the animal moves through vegetation. These casts are sufficient to identify the species from the scale patterns.

Another important feature of skin is its colouration, with functions including camouflage (most species), warning of toxicity or danger to potential predators (great crested newts, adders) or enhancement of status during courtship (male sand lizards).

Fig. 2.4 Shed reptile skin (Keri Leaman)

2.3 Feeding

Post-metamorphic amphibians and lizards of all ages are skilled at catching invertebrates although techniques and dietary preferences vary considerably. All amphibians swallow their prey whole and it is surprising how large a creature can be accommodated. On the other hand they are not good at assessing size and regularly give up on items that, after a struggle, turn out too big to cope with. Newts rely on smell and vision, approaching prey cautiously and seizing it in jaws equipped with tiny retaining teeth. Small worms, slugs and arthropods all feature in the diet when on land. In the breeding ponds, Daphnia, water hoglice (*Asellus*), leeches, caddis larvae and frog tadpoles are favoured. Great crested newts, but not the other species, also eat the generally unpalatable toad tadpoles. Newts are much more likely than frogs or toads to attack immobile prey, probably because of their better sense of smell, especially if the victim is already injured. Frogs, especially water frogs, leap after flying insects with mouth agape. Usually, though, potential prey is approached to within a couple of centimetres, caught by a rapid flick of an extensible, sticky tongue and returned to the mouth within milliseconds. The victim has to move to attract attention, and staying still is an excellent defence against frogs and toads which rely heavily on seeing motion to trigger an attack. Unlike the newts, none of our native anurans feed underwater. On land, common frogs prefer slugs, snails and worms whereas common and natterjack toads go more for arthropods, especially beetles, moths and flying ants. Natterjacks feed around the ponds during their protracted breeding season but common frogs and toads, in their briefer and more frenetic spawning period, usually show no interest in food until they leave the water.

Reptiles are much better than amphibians at assessing whether potential prey is of a manageable size. Viviparous and sand lizards actively chase spiders, moths, butterflies and any other arthropods (even bees) that come their way before seizing them in their jaws, giving a shake to stun if necessary, chewing for a while and then swallowing them whole. Slow-worms are much less agile but in light of their dietary preferences this is of no consequence. They approach potential food, usually small slugs, snails or worms, with calculated deliberation before grasping decisively in a manner not unlike that of newts living on land.

Snakes have to contend with animals that are often faster than they are, and which in some cases can fight back and inflict serious damage. Strategies vary but it's well known that snakes can swallow prey much larger than looks possible from the size of their head, due to elastic and flexible connections between various jaw bones. Grass snakes are excellent swimmers and chase amphibians, their prey of choice, both on land and in the water. Frogs caught in a pond are brought to the bank for consumption. Adders can hunt actively but often adopt a 'sit and wait' approach, lunging at a mouse or vole when it passes close enough, injecting venom and giving time for it to work before seeking out the dead or dying victim (by smell, using the tongue to pick up scent). Small mammals are high on their list but adders also take fledgling birds, other reptiles and occasionally amphibians. Smooth snakes probably hunt underground in sandy tunnels and burrows for much of the time, though this is hard to observe. Other reptiles, including snakes, are their main prey which they often capture by throwing coils around the victim to assist with restraint though they do not constrict to asphyxiate as pythons do. All snakes swallow their prey whole. Because the meals are large, feeding is intermittent and eating just once a month can suffice to maintain body condition.

2.4 Respiration

It is not only in the range of breeding behaviours that amphibians show greater variety than other vertebrates. Respiration can occur via three separate surfaces, notably lungs, the floor of the mouth (buccal cavity) and across the entire skin. Frogs, toads and newts pulsate their throats continuously while on land, taking in air through the nostrils and making use of the buccal cavity, richly supplied with blood vessels, for gas exchange. Occasionally deeper throat movements occur, transferring air into the lungs with the nostrils closed. This cannot be done by chest expansion because amphibians lack a complete rib cage. If nostrils become blocked, as often happens when toads are infected by larvae of the fly *Lucilia bufonivora*, breathing is severely impeded and the animals show great distress because amphibians cannot breathe through an open mouth. Death is the inevitable result. Efficient respiration absolutely requires a 'closed system' between mouth and lungs. Species vary considerably in their relative use of the three surfaces. Toad skin is relatively dry and impermeable and for them the buccal cavity and lungs predominate in respiration. Frogs, with damper skins, can survive for long periods breathing through the skin alone. In winter they frequently rely on this during hibernation at the bottom of ponds. Toads almost always overwinter on land but can survive in highly oxygenated water, occasionally hibernating on the bottom of streams. Some amphibians, notably pletho-dontid salamanders of the Appalachian mountains, do without lungs altogether. Larvae of all species have gills, internal and thus inconspicuous in the case of frog and toad tadpoles but evident as large, feathery structures behind the eyes in newt tadpoles. Again there are some important differences between species. Frog tadpoles develop lungs quite early in development and if ponds become anoxic can survive by gulping in air from the surface. Toad lungs appear much closer to metamor-phosis, making their tadpoles more vulnerable to oxygen depletion.

Reptile respiration is conventional and in all British species relies solely on lungs. In snakes, however, body elongation has led to most species having only a single functional lung. Some turtles have vascularised skin patches within their cloacae and mouth cavities that can support respiration underwater for long periods, allowing them to hibernate at the bottom of ponds just like frogs.

Vascularised
Supplied with blood vessels

2.5 Thermoregulation

As ectotherms unable to generate their own body heat, all amphibians and reptiles sustain core temperatures by interacting with their environment. In the case of amphibians this mostly involves soaking up background infrared radiation at night when they are hunting from spring through to autumn. Some, however, are more proactive and water frogs bask extensively in bright sunshine to raise and regulate their body temperatures. In reptiles thermoregulation is more sophisticated and extensive. Lizards and snakes of all the British species spend time basking, either overtly or covertly, to raise their temperatures high enough to operate effectively. In spring and autumn this can happen at any time of day to compensate for fluctuating and often low environ-mental temperatures. In summer there is less need for it and basking mostly occurs just briefly at the start and end of the day. Female adders are an exception to this

Thermoregulation
Active control of body temperature

rule, basking extensively in August prior to giving birth. It is this basking behaviour that renders the animals most visible to human observers and predators alike, and so it is usually carried out close to cover into which a rapid retreat is always possible. In early spring some species, such as adders and sand lizards, flatten their bodies while basking to maximise exposure and thus speed up the warming process. This is particularly important for males which are maturing sperm in their testes immediately after emergence from hibernation.

2.6 Amphibian reproduction

The breeding behaviours of amphibians and reptiles are wonderfully varied and fascinating to watch. Common frogs and toads are 'explosive' breeders, completing the process within a week or two unless interrupted by cold weather. They arrive at the ponds early in spring, frogs usually a little before toads, after migrating from their hibernation sites. Toads are well known for these journeys which may be several kilometres long and often involve hundreds or thousands of individuals. Both species frequently spawn in the same pond or ditch but frogs are more catholic in their choice than toads, and in any particular area frogs usually breed in more places. Activity is mostly nocturnal, but can sometimes be watched in daytime too, especially in large populations. Male frogs congregate around a shallow area where females progressively generate a 'spawn mat' together, a mass of spawn in which individual females' contributions are not distinct. Male frogs croak with a quiet purring sound, and pairs in 'amplexus' (the mating embrace) move among them at the spawn site. Fertilisation is external, males shedding sperm in synchrony with egg deposition. Egg laying by any particular female is completed within seconds, usually at night, after which the pair separates. Spent females leave the spawning area immediately. The newly-deposited spawn, initially no bigger than a walnut, absorbs water very quickly and expands to form the familiar jelly-like mass with many hundreds of eggs.

Common toads do things a little differently. Males greatly outnumber females and fight extensively for the right to mate. Often males spread out around the pond to ambush females and some pair up before they even reach the pond. The contests are essentially wrestling matches. Sometimes many males latch on to the same

Fig. 2.5 A mating ball of common toads *Bufo bufo* (Quentin Scouflaire)

female (**Fig. 2.5**), a dangerous situation for her because if the situation persists she may drown. As with frogs, spawning focuses on one or a few places in a pond but in deeper water than frogs prefer. The process is protracted over several hours because the hundreds or thousands of eggs are laid in a long string with irregular bouts of emission and concurrent fertilisation. Males have a quiet croak, mainly used by individuals at the edge of the colony to try and attract any late-coming females.

By contrast, natterjack toads and water frogs are 'protracted' breeders. They spawn later in spring or in early summer and males hang around the breeding sites for several weeks. Females visit only briefly to lay their eggs but individuals can arrive at any time over the breeding period so spawning is usually much less synchronous than that of common frogs or toads. Male natterjacks and water frogs have loud and distinctive calls to attract females. Natterjacks generally choose temporary ponds that will dry up later in summer, whereas water frogs use deep ponds or ditches for reproduction. Unlike common toads, in which females congregate and intertwine their spawn strings together, natterjack females lay theirs individually and in much shallower water, making them relatively easy to find. Water frog females lay several small spawn clumps rather than the single large one produced by common frogs, and hide them in weed making them hard to spot. Unlike the other species, water frogs commonly breed by day as well as at night and at the height of the season there is frenetic activity with much calling and males chasing each other around the pond.

All our native newts are 'protracted' breeders, spending several weeks in the ponds after arriving from their hibernation sites. Reproductive behaviour is broadly similar for all of them. Unlike anurans they have no amplexus embrace and little direct contact of any kind between the sexes. Males make impressive displays, chasing females and trying to get in front to block their passage. If successful (that is, if the female doesn't swim away) this gambit is followed by a series of movements in which the male's tail is lashed vigorously against his body, wafting his scent to the attending female (**Fig. 2.6**). If she remains interested (often she doesn't) he moves around and deposits sperm in a gelatinous, white sack – the spermatophore – on the pond bottom. She then moves forward to position her cloaca over it and take up the sperm. Fertilisation is subsequently internal. The female then lays her two or three hundred eggs (many fewer than frogs or toads) individually, wrapping them wherever possible in the leaves of pond weeds to provide protection from predators. Male newts develop crests in the breeding season although these are very low and

Amplexus embrace

Grasping of a female by a male frog or toad during the breeding season

Fig. 2.6 Smooth newt courtship (Charles Snell)

unimpressive in palmates and almost nonexistent in alpines. Great crested newt females mate preferentially with males that have the largest crests. Mating and spawning are not highly focused in a particular spot but occur at many places in a typical pond wherever there is suitable weed, although crested newts have a crude lekking system in which males select an area on the pond floor and defend it by trying to chase away same-sex intruders. This can be best observed after dark using a powerful torch. Most mating and spawning by all species occur at night but usually there is something to be seen, especially with smooth newts, in daytime when breeding is at its peak in mid spring.

Lekking
The tendency of males to assemble at a communal mating place; the display behaviour of males at that site

2.7 Reptile reproduction

Like amphibians, all British reptiles mate in spring. Unlike amphibians, however, egg laying or the production of live young occurs after a substantial interval. Male reptiles tend to emerge from hibernation before females and bask extensively to mature their sperm. In the case of sand lizards the difference in emergence time between the sexes can be a month or more. By the time females appear, males are in breeding condition and ready to mate. They have shed their skin and colours are at their brightest. On first emergence, often in March, male sand lizards are a dull grey or brown but by April their flanks are glorious iridescent green. Mating is a highly competitive affair. Males of all British lizards, including slow-worms, fight and chase other males to win access to females. These fights can be serious and many old males bear scars of previous engagements around the head and neck. Tails, which can be shed by all British lizards to deflect the attention of predators (in a process known as autotomy), are also regularly lost by internecine strife in spring. Size matters and it is usually the largest males that do best. Dominant male sand lizards mate with females following a rough courtship which involves biting her and holding her in place while the penis is inserted into her cloaca. Pairing can last from minutes to at least half an hour. Much the same procedure is followed by the other lizard species. Male sand lizards attempt to mate-guard after copulation, though they're not very good at it, and pairs can often be seen basking together.

Mate guarding
The guarding of a receptive female by a male, to prevent other males from mating with her

Female sand lizards lay one or two batches of eggs, depending on the weather, usually between late May and

Fig. 2.7 Pair of male adders fighting for dominance (Richard Newton)

early July. A typical clutch consists of five to ten eggs, laid in a hole dug by the female at night. Hatchling lizards appear any time from late July through to September. Viviparous lizard females deposit their live young, typically seven or eight, in a secluded spot in vegetation some time during late June or July. Slow-worms are usually the last to finish giving birth, with clutch sizes of up to more than twenty (but usually far fewer); live young are deposited in August or September.

Snake reproduction follows a broadly similar pattern to that of lizards. Adders are well known for communal hibernation at especially favourable places, such as burrows under old trees, and sometimes many can be seen basking together after they emerge on sunny days in early spring. By April males have shed their skins and the black zig-zag back patterns contrast sharply with the silvery backrounds, a much sharper definition than occurs in females. Again competition is fierce and the snakes become relatively bold. This makes them more vulnerable than normal to predators, but is convenient for interested human observers. Males chase each other with bodies raised and intertwined in a wrestling match commonly known as the 'dance of the adders' (**Fig. 2.7**).

Eventually one will triumph and drive the other away, leaving him free to mate with a female normally not far distant. Again there is interwining of bodies, and penis insertion for sperm transfer. Grass snakes and smooth snakes have similar mating behaviours although they are less often witnessed. 'Balls' of grass snakes involve many males all wrestling around a female, attempting copulation. Both of these species mate later than adders, mainly in May, though the act has very rarely been seen in the wild in the case of the highly secretive smooth snakes.

Adders give birth to an average of about nine live young in August or September. Smooth snakes do much the same, in September or sometimes October. Females of both these species do not breed every year but typically once every two or even three years. This delay is believed to reflect the time it takes to accumulate the large amount of resources needed to sustain a clutch of developing snakes from fertilisation through to live birth. Grass snakes lay very variable numbers of eggs (from five to more than twenty) in June or July, every year as far as we know, in sites such as compost heaps that generate heat, which speeds development. Young snakes hatch several weeks later, in August or September.

Most of our native amphibians and reptiles exhibit multiple paternity. Sperm shed at frog and toad breeding sites, sometimes by unpaired males, can fertilise more than one spawn mass. Female newts regularly take up spermatophores from several suitors. Male reptiles often attempt mate-guarding but it is rarely completely successful and female lizards and snakes usually pair with more than one male. Pairs of natterjacks commonly spawn in isolation and may be an exception. It is clear though that multiple paternity is a sensible strategy for females because it usually results in more viable and fitter offspring.

2.8 Taxonomy and evolutionary history

The taxonomy of amphibians and reptiles has undergone a transformation recently following a period of relative stability through much of the twentieth century. The new changes are mostly due to the application of molecular methods rather than just the morphological ones (body size, internal anatomy, shape and colour) used previously. More than 6,000 species of amphibians are now recognised globally, an increase of 50% within the last twenty years. This has come about partly by the discovery of species previously unknown to science, but also as a result of molecular analyses revealing cryptic species (which are visually indistinguishable) within those previously considered to be one. With respect

Molecular methods
Use of molecular data, usually DNA sequences, to infer relatedness and population structure

Cryptic species
Species which look the same but don't interbreed and are therefore distinct

Table 2.1 Nomenclature of amphibians and reptiles in Britain, excluding rare escapes. Where alternative common names exist, those in bold are used throughout this book.

Common name(s)	Current scientific name	Previous scientific name
AMPHIBIA		
Frogs & toads (Anurans) - native		
Common frog	*Rana temporaria*	
Pool frog	*Pelophylax lessonae*	*Rana lessonae*
Common toad	*Bufo bufo*	
Natterjack toad	*Bufo calamita*	*Epidalea calamita*[1]
Frogs & toads - introduced		
Marsh frog	*Pelophylax ridibundus*	*Rana ridibunda*
Edible frog	*Pelophylax kl. esculentus*[2]	*Rana esculenta*
Midwife toad	*Alytes obstetricans*	
African clawed **frog**/toad	*Xenopus laevis*	
Newts (Urodeles) - native		
Smooth or common newt	*Lissotriton vulgaris*	*Triturus vulgaris*
Palmate newt	*Lissotriton helveticus*	*Triturus helveticus*
Great crested or warty newt	*Triturus cristatus*	
Newts - introduced		
Alpine newt	*Ichthyosaura alpestris*	*Triturus alpestris; Mesotriton alpestris*[3]
Italian crested newt	*Triturus carnifex*	*Triturus cristatus*[4]
REPTILIA		
Lizards - native		
Viviparous or common lizard	*Zootoca vivipara*	*Lacerta vivipara*
Sand lizard	*Lacerta agilis*	
Slow-worm	*Anguis fragilis*	
Lizards - introduced		
Wall lizard	*Podarcis muralis*	*Lacerta muralis*
Western **green** lizard	*Lacerta bilineata*	*Lacerta viridis*[5]
Snakes - native		
Adder or Viper	*Vipera berus*	
Grass or Ringed snake	*Natrix natrix*	
Smooth snake	*Coronella austriaca*	
Snakes - introduced		
Aesculapian snake	*Zamenis longissimus*	*Elaphe longissima*

[1] A change to *Epidalea* was recently proposed but then rejected, at least for now. [2] The edible frog is a fertile hybrid of marsh and pool frogs, resulting in a bizarre form of hybridogenetic reproduction (hence kl. = klepton). Details of the mechanism are given in Beebee and Griffiths (2000). In Britain edible frogs normally occur only in mixed populations with (introduced) pool frogs. [3] Alpine newt nomenclature has undergone a rapid recent transition at genus level, from *Triturus* through *Mesotriton* to *Ichthyosaura*. [4] Italian crested newts were formerly classified as a subspecies of *cristatus*. [5] All European green lizards were, until recently, classified as one species (*viridis*).

to the British amphibians, however, the main result of recent taxonomic revisions has been an increase in the number of recognised genera but no change in the number of species. For the reptiles, with around 9,000 species globally, much the same is true, but in general classification changes relevant to the British reptile fauna have been smaller than those relating to our amphibians. The current nomenclature of the species found in Britain, together with some alternative and/or earlier names, is summarised in **Table 2.1**.

Native pool frogs, midwife toads, African clawed frogs, green lizards and Aesculapian snakes are very rare in Britain and each persists in just one or a very few populations. American bullfrogs *Lithobates catesbeianus* (formerly *Rana catesbeiana*) established themselves at several places in recent years but have been, or are being, exterminated as fast as possible.

The earliest amphibian fossils date back some 350 million years, to the time when vertebrates were first able to survive out of water. Current amphibians, the 'Lissamphibia', comprise three divergent orders: Anura or Salientia (frogs and toads), by far the most diverse group today with more than 5,000 species; Urodela or Caudata (newts and salamanders), with a few hundred species; and the Caecilia or Gymnophiona, worm-like burrowing animals (**Fig. 2.8**) confined to the tropics, with fewer than two hundred species. The ancestors of modern anurans and urodeles split from those of caecilians some 330 million years ago, and anurans and urodeles separated later, probably around 260–270 million years ago. Anurans occur today on all continents except Antarctica but, curiously, urodeles are predominantly animals of north temperate countries with very few species in the tropics. Amphibian evolutionary relationships, summarised in **Fig. 2.9**, are based increasingly on molecular data but are still in a state of flux with some tree branches not fully resolved.

Fig. 2.8 A caecilian *Dermophis mexicanus* (Franco Andreone)

The Bufonidae and Ranidae are two of the most biodiverse families but evolutionary classification confirms, as shown in **Fig. 2.9**, that there is no fundamental distinction between 'frogs' and 'toads'. These names are historical relics and 'tree frogs', for example, are much more closely related to bufonid 'toads' than to ranid 'frogs'.

Reptiles evolved from early amphibians about 310 million years ago. A comparable, simplified tree of evolutionary relationships among extant reptiles is shown

Fig. 2.9 Evolutionary relationships of amphibian families. A simplified version reprinted from: Pyron, R.A. and Wiens, J.J. (2011) A large-scale phylogeny of Amphibia including over 2800 species, and a revised classification of extant frogs, salamanders, and caecilians, *Molecular Phylogenetics & Evolution* **61**, 543–583; with permission from Elsevier. Dashed lines represent one or a few minor families with generally very few species. Genera found in Britain are boxed.

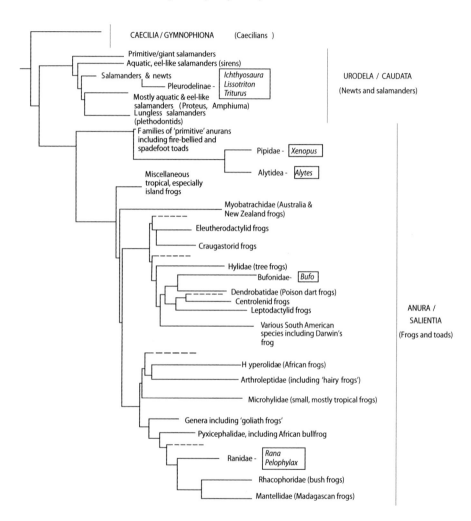

in **Fig. 2.10**, based once again on morphological and molecular data (Hedges and Poling, 1999). A big recent surprise was the discovery that reptiles are not monophyletic, meaning that they do not cluster as a single discrete group, the sole descendants of a single common ancestor. Archosaurs, including crocodiles and turtles, are more closely related to birds (which evolved from dinosaurs,

Fig. 2.10 Evolutionary relationships among reptile families (after Hedges and Poling, 1999). Genera found in Britain are boxed.

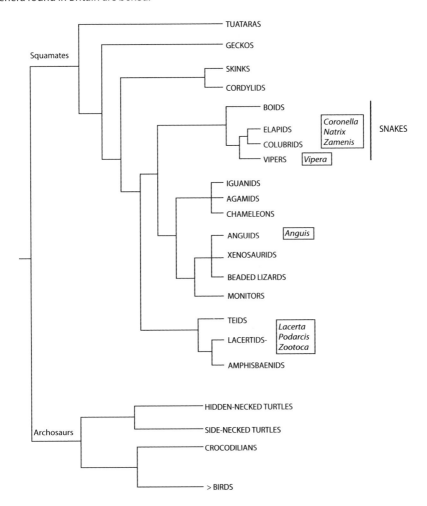

an extinct archosaur group) than to squamates which include snakes (ophidians), lizards (saurians) and the distinctive New Zealand tuataras (rhynocephalians). No terrestrial turtles (tortoises) or terrapins of the ancient Chelonia group are native to Britain although escaped specimens of various species turn up occasionally and one marine turtle, the leatherback, is a regular summer visitor to British seas.

3 Ecology and conservation

Ecology is a wide-ranging discipline that includes the study of animal behaviour, population dynamics and community structure. This chapter focuses on aspects particularly relevant to naturalists and to practical conservation work. Reviews of amphibian and reptile ecology are provided by Wells (2007), Reilly, McBrayer and Miles (2007) and Mullin and Seigel (2009). In the context of contributions to research accessible to amateur study and of benefit to conservation, population ecology is arguably the most important component of the subject and therefore a primary consideration here.

3.1 What is a population?
The concept of population is central to understanding how animals and plants survive in the wild. Unfortunately the definition of a population is as difficult as the famously problematic definition of species. Where does one population end and another begin? Attempts to answer this question are important for many kinds of study. There is no universally agreed solution but the situation is often clearer in amphibians and reptiles than in many other taxonomic groups. This is because herpetofauna (the collective term for amphibians and reptiles) are rather sedentary and throughout their lives most individuals do not wander more than a few kilometres from their place of birth. In many cases the distances are far shorter, just hundreds of metres. This means it is sometimes possible to identify a group of individuals that constitute, at least approximately, a discrete population. Contrast this with flying animals such as many insects, birds and bats, or plants with seeds and pollen carried across great distances. High individual mobility inevitably blurs boundaries between populations. This problem is widespread. Genetic studies have indicated that badgers in Britain, from Cornwall to Scotland, cannot be subdivided sensibly into anything less than one continuous population.

In the case of amphibians and reptiles the difficulty in defining populations varies among species. Generally it is easiest with the rarest. Some localities with natterjack toads, sand lizards and smooth snakes occur as small 'islands' of suitable habitat widely separated from each other by terrain the animals will rarely if ever cross.

Gene flow

Transfer of genetic
material between or
within populations, e.g.
by migrating animals that
breed on arrival

In these cases the assumption of a discrete population seems reasonable but even with the rarities it is not always straightforward. Many blocks of heathland in Dorset are separated by short distances and by habitats not completely hostile to sand lizards or smooth snakes. The same applies to dunes and marshes used by natterjacks around the Irish Sea coast. Ideally to define a population we want to know whether migration occurs between different places where we find the animals, but this information is difficult to get if long-distance movements are rare and/or mainly carried out by juveniles, as is often the case. Genetic analysis offers a powerful way of assessing gene flow indirectly but is expensive and for the British species has only been done comprehensively in the case of natterjacks (Rowe and Beebee, 2007). Genetic studies have confirmed the difficulties of treating populations as discrete entities completely closed to immigration and emigration. However they do permit informed estimation of which populations are relatively distinct with only occasional movement in or out, and in most cases that is good enough.

The situation with the widespread species is even more challenging, and the most abundant such as common frogs probably, like badgers, constitute single populations across large areas of Britain. A widespread assumption in the past was that, for amphibians, a pond was equivalent to a population. Unfortunately in most cases this is too simplistic. Should we expect sand dune pools just a few metres apart to support distinct natterjack toad populations? Of course not, but the same reservation often applies to more widely separated ponds in the countryside at large or, in the case of reptiles, to banks of good habitat connected by open forest rides or hedgerows. Many species exist as 'metapopulations': with clusters of breeding sites in which the subpopulations experience occasional local extinctions rescued by subsequent immigration, and with some sites ('sources') faring much better than others ('sinks') which would not persist in the absence of these movements (Griffiths, Sewell and McCrea, 2010).

How does all this affect studies of amphibians and reptiles? For some investigations, such as those into behaviour, the definition of population is usually irrelevant and can be ignored. For others, particularly those concerned with long-term persistence and conservation, the issue must be addressed. However in practice

Inbreeding depression

Loss of individual fitness due to accumulation of damaging mutations as a result of breeding between related individuals

Philopatric

Tending to return to breed at the site of birth

it need not be too daunting. It's usually possible to decide whether the study site is likely to be closed to migration (that is, a discrete population) or open to it (so probably a subpopulation) and then analyse results accordingly. Ironically, although discrete populations are attractively simple for study they are often the least desirable from the animals' perspective. Prolonged isolation, especially if the population is small, generates a high risk of inbreeding depression and of extinction from random factors such as a few bad winters with no prospect of recovery through immigration.

Two kinds of information can help us decide what kind of population exists in a specific place. The first relates to the habits of the species in question. Common toads are highly philopatric, meaning that most return to breed at their pond of birth, and they also tend to use relatively few ponds in a given area. So although adults wander up to several kilometres it is often safe to assume that a common toad breeding pond represents a largely closed population. By contrast, common frogs and grass snakes move across a wide range of habitats and for them it is probably safest to presume subpopulation status in most situations. With newts, which are generally less mobile than frogs and toads, decisions benefit from greater emphasis on the second kind of information, notably habitat features. Multiple ponds not too far apart imply a metapopulation structure if terrestrial habitat between them is favourable (that is, not intensively arable or intersected by major roads). By contrast, isolated ponds in poor terrestrial habitat may sustain closed populations. The same might well be true of adders and common lizards on discrete patches of ideal habitat separated by intensive agriculture. This logic applies to all species superimposed on their inherently different tendencies to move around. Gardens are interesting in this context because of their value for amateur studies. In most cases the animals encountered in gardens, be they amphibians using a pond or slow-worms in rough grass, will almost certainly be subpopulations mixing with neighbours and should be treated as such.

3.2 Population dynamics

An understanding of population dynamics, especially sizes and age classes, is important for estimating long-term viability and therefore also for conservation. It is rarely easy to quantify population dynamics, but many amphibians and reptiles offer reasonable prospects for doing this because of their visibility in predictable places, be they ponds or basking areas, during the spring breeding seasons. Estimates of total population size (censuses) are an obvious starting point. Amphibians or their spawn can be counted in breeding ponds, and reptiles can be registered at basking sites along transects through suitable habitat, or under refugia, using methods described in chapter 6. But what do the numbers obtained in this way really mean? Some species are more amenable to census counting than others. Common frogs and natterjacks are the most straightforward because in both cases numbers of spawn clumps or strings can be counted accurately. However, female natterjacks do not necessarily breed every year, depending on their body condition and on weather in a particular season. Spawn counts for both these most amenable species can only generate approximations of total adult population size, inevitably biased by variable female behaviour and by the need to assume a particular sex ratio. Reptiles are at the other extreme with respect to interpreting count data, as only an unknown fraction of the population will be seen basking along a transect. Despite these caveats it is usually possible to obtain useful information just by counting, though more thorough investigations employ capture-mark-recapture (CMR) methods based on the Lincoln Index. In this procedure individuals are caught, given a mark to record the event (such as by clipping a toe or injecting a PIT tag) or photographed as described in chapter 1, and released, and another sample is caught again some time later. The population size P can then be estimated as:

$$P = \frac{Nm_1 \times Nm_2}{N}$$

Where Nm_1 = number of animals recorded during first capture round; Nm_2 = number of animals caught in second capture round; and N = number of animals recorded in first round and recaptured in the second. Multiple recapture rounds increase the accuracy of the estimate of P, and software packages such as MARK (available as freeware from www.phidot.org/software/mark/) carry out these calculations allowing various model assumptions such as whether or not the population is closed to migration. In the case of

Transect
Line drawn across an area of land for the purpose of conducting a survey

Sex ratio
Numbers of males relative to numbers of females

amphibians survival of larvae through to metamorphosis can also be estimated by CMR techniques, although it's not easy to mark tadpoles! CMR obviously needs a lot more effort than head or spawn counting, may require licences (see **9.1**) and is not always straightforward, especially with reptiles. Basking sites are often used by the same individual time after time, but Lincoln Index methods assume random mixing across the population between the sampling operations. Wheater and Cook (2003) give more detail.

Second in importance after population size is demography, an understanding of age structure and recruitment rates. In this field too, many amphibians and reptiles (apart from slow-worms and snakes) are more amenable to study than most other vertebrates. This is because their long bones accumulate annual rings, just like trees, when growth ceases during each hibernation period. By cutting off a toe tip and staining the peripheral bone ('skeletochronology'), it is possible to make an accurate estimate of an individual's age without sacrificing the animal (**Fig. 3.1**). Sampling across a population thus provides an immediate snapshot of its age structure. Nothing comparable is possible

Recruitment
Addition of new members
to a population

Fig. 3.1 Age structure
of natterjacks on the
Merseyside dunes
as determined by
skeletochronology (Denton
and Beebee, 1993).

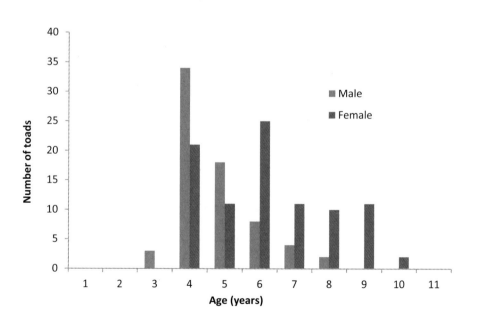

for birds or mammals, though fish ages can often be determined from scale marks. The only alternative, marking animals and measuring individual survival rates over many years, requires much more time and effort. Demographic studies have been carried out on all the British amphibians as well as on viviparous and sand lizards, though not always on British populations. By contrast we have little robust information on the age structure of snake populations, an exception being smooth snakes in Wareham Forest which were subject to a long-term CMR study (Reading, 2004).

Taken together, data on population size and age structure provide valuable insights into population health and viability. Broadly speaking bigger is better, ideally with evidence of continuous and/or recent recruitment and a broad range of age classes. Genetic analysis using molecular techniques such as micro-satellite genotyping adds to this picture by allowing calculation of the 'effective' population size (Beebee, 2005) and by indicating the degree of gene flow and thus isolation, or otherwise, of the population. Effective population size is the number of animals successfully reproducing in each generation, and is often far smaller than the census size (many males never succeed!), but critically important for long-term viability.

3.3 Population trends

For conservation purposes it is important to know not just the basic features of a population but also whether it is stable, increasing or decreasing. For this there is no substitute for observations extended over many years, repeating assessments of size and age structure at regular intervals. Naturally it is declines that cause most concern but identifying these and their causes can prove very challenging. Is it possible to assess population trends at a national scale? Britain has a long history of wildlife recording, and data on amphibians and reptiles have been published on several occasions since 1947. However these early efforts were not systematic and not designed for quantifying national trends. The data are confused by greatly increased recording efforts over the past fifty years, which often imply increases even in periods where specific studies showed the opposite (for example, declines of common frogs in the 1960s and 1970s). The temptation to make inferences from such imperfect data should therefore be resisted (Beebee, Wilkinson and Buckley, 2009).

Potentially more useful are protracted studies of individual populations and there are a few such datasets extending up to thirty years or more. However these often highlight another difficulty with assessing long-term trends, notably the propensity of populations, especially those of amphibians, to fluctuate wildly over periods of a few years. These variations, presumably driven by natural cycles of some sort, can span several orders of magnitude. For some species snapshots across a few years are more likely to show declines than stability or increase, even in healthy populations. This is because they experience occasional 'boom' years of very high recruitment and subsequent rapid increase separated by a succession of poor breeding seasons with consequent gradual declines. As Meyer, Schmidt and Grossenbacher (1998) discovered, common frogs in three Swiss ponds were apparently declining most of the time, but in only one case (where fish had been introduced) was there an underlying long-term decrease in numbers.

All of this means that detection of trends on either national or local scales requires carefully organised efforts extending over many years. The problems are not insurmountable, though, and recent approaches are discussed in chapter 4 including ways in which amateur naturalists can make worthwhile contributions. Whether we are dealing with short-term oscillation or a long-term trend, it is interesting to try and understand what forces drive the changes observed over time in wild amphibian and reptile populations. There are several candidate mechanisms.

3.4 Natural regulators of population size

Factors usually considered 'natural' include competition, predation, disease and environmental variation. Some animal populations oscillate over time in cyclic patterns based on changes in predator or prey abundance, or on the arrival of newly mutated pathogens later defeated by immune responses. To what extent can these factors account for the fluctuating sizes of British amphibian and reptile populations? In most cases we simply don't know the answer. Does prey abundance regulate amphibian or reptile populations by intensifying competition for limited resources? Tadpoles at high densities grow very slowly, mainly due to food limitation, and can experience high mortality rates especially if they fail to metamorphose before winter sets in. Survivorship during the

Pathogen
Harmful microorganism

larval phase may be an important determinant of adult population size and, at least for natterjacks and great crested newts, adult numbers often correlate with survival to metamorphosis two or three years previously. Survivorship from egg to metamorphosis is normally very low, typically just a few percent on average, for most amphibian species. However this can vary enormously between years and localities, from zero to 50% or more, so there is obviously scope for regulating population size in this phase of the life cycle.

It is likely that juvenile mortality can also be a substantial driver of population dynamics. Froglets and toadlets emerging in their thousands around midsummer may be at high risk of starvation although evidence on this is anecdotal. Young amphibians and reptiles need plenty of food to sustain rapid growth, which if achieved reduces vulnerability to predation. Adult amphibians and reptiles can survive for long periods without food but prey scarcity could have long-term effects on winter survival or subsequent fecundity, which in turn might trigger population fluctuations. Body condition (weight per unit length) varied inversely with respect to population density in adult natterjacks across Britain (Denton and Beebee, 1993) which might imply a regulatory role of intraspecific competition for food.

Interspecific competition, on the other hand, seems to be rare among the British amphibians and reptiles, in accord with ecological theory which predicts that species survive by establishing their own distinct niches in an ecosystem. But despite this expectation smooth and palmate newts often breed in the same pond and exhibit very high niche overlaps (for example in microhabitat selection, or prey) while coexisting in substantial numbers (Griffiths, 1986). This probably means that the populations of both species are regulated by some other factor, such as predation or environmental variation, so there is no damaging competition for resources. However, disturbance can have profound effects. Changes in habitat structure (scrub encroachment with increased shade and humidity after cessation of grazing) led to invasion by common toads of many ponds on heaths and dunes, habitats previously hostile to them because of high temperatures and aridity. This in turn allowed them to outcompete and locally exterminate natterjack toads because the common species' tadpoles hatch early

Fecundity
Numbers of eggs produced per female

Intraspecific
Within a species
Interspecific
Between species

in spring and devour any natterjack spawn laid later, or inhibit the growth of natterjack tadpoles so that these fail to metamorphose before the temporary ponds dry up in midsummer (Bardsley and Beebee, 1998).

Amphibians and reptiles are consumed by a wide range of predators. Frogs, toads and newts are taken by, among others, pike, herons, crows, gulls, buzzards, rats, hedgehogs, mink, otters, foxes, badgers and grass snakes. The complete list including rarer predation events would be a long one. Despite their skin toxins, annual survivorship of toads is little better than that of frogs. Frog spawn is eaten by ducks, moorhens, tadpoles, newts and flatworms. Newt eggs are consumed by other newts and by insect larvae. Frog and newt tadpoles (and sometimes adults, **Fig. 3.2**) fall prey to fish, newts, dragonfly and water beetle larvae and bugs such as water boatmen and water scorpions. Toad tadpoles are distasteful to most vertebrates and survive well in fish ponds but are consumed avidly by many invertebrates.

Lizards provide food for birds of prey, especially kestrels, opportunistic mammals and snakes. Domestic cats and blackbirds are effective predators in urban areas where they exist at high densities, probably accounting for the scarcity of viviparous lizards in gardens. Cats also

Fig. 3.2 A diving beetle larva eating a smooth newt (Julian Whitehurst)

catch the more secretive slow-worms but seemingly not often enough to have big effects on their populations. Adders take lizards and the smooth snake specialises in them. Snakes are preyed upon by buzzards, mammals (hedgehogs seem particularly fond) and no doubt other species not often witnessed making an attack.

Again the question arises as to whether predation materially affects amphibian or reptile population dynamics. For lizards in gardens this may be true but we have no hard data. Most predators are opportunistic and their abundance seems unlikely to cycle in a manner dependent on amphibian or reptile numbers. Grass snakes and smooth snakes, as amphibian and reptile specialists respectively, might be more closely linked to their prey but there are few definitive studies. The fecundity of adders correlated with cycles of vole abundance on a Swedish island but the isolation of this snake population, with limited food choice, may mean it was not a typical case (Forsman and Lindell, 1997). The best evidence of predation power is that of fish on great crested newts (Baker, 1999). Introducing fish to a pond, even small species such as sticklebacks, can exterminate great crested newts within a few years because their larvae, swimming in open water, are highly vulnerable to attack. Again we have a situation where artificial alteration of the habitat is of prime importance because this usually happens when humans, especially anglers, introduce fish into ponds where they were previously absent.

Amphibians and reptiles have a range of microbes and parasites associated with them but in most cases these do not generate detectable symptoms. There may be longstanding host-pathogen cycles but none have been identified. There are widespread reports of mass mortalities of amphibian spawn, and tadpoles disappearing for no obvious reason, and these could be caused by pathogens. However, two amphibian diseases are of unquestionable significance. Both seem to be emergent, meaning they probably arrived in Britain recently and susceptible species had no previous experience of them. This is often a recipe for disaster, as with myxomatosis and rabbits in the 1950s. *Ranavirus* mostly infects common frogs though common toads can also die from it. The pathogen has spread extensively especially in southern and eastern England and in garden environments. Eighty percent or more of a frog population is often wiped out within weeks and recovery can take many years, though it usually happens eventually (Teacher, Cunningham and Garner, 2010). Although epidemics are locally devastating and gruesome to see, with emaciated animals dying conspicuously mainly in the summer months, there is no evidence of an overall decline of frogs in areas hit by this disease. The other newcomer, chytrid fungus *Batrachochytrium dendrobatidis* ('Bd'), was discovered in the late 1990s. It has had massive impacts in some tropical countries where many species have declined hugely or even become extinct as a result of chytrid infection. Bd has been present in Britain since at least 2004 and within a few years most natterjack toad populations became heavily infected. Fortunately at least so far this fungus has not been associated with widespread disease symptoms or population declines. The results of Bd infection are highly variable among species and also relate to local climatic conditions. Infection without catastrophic disease seems quite common and we can only hope that Britain's amphibians are similarly favoured. Once again a longstanding natural system (in this case the host-pathogen relationship) has been altered by humans, the most likely importers, albeit acciden-

Fig. 3.3 Results of a heath fire in Dorset (Nick Moulton)

tally, of these new threats. Nothing comparable has yet happened with reptiles. Lizards often carry large burdens of blood-sucking ticks and some may die as a result but this seems unlikely to influence population dynamics.

Last but not least among factors affecting population numbers is environmental variation, such as changes in temperature or rainfall that influence recruitment or survivorship. Dry springs causing early desiccation of breeding ponds can eliminate breeding success of natterjack toads and sometimes of other species. Perhaps surprisingly, wet winters increase mortality in great crested newts (Griffiths, Sewell and MacCrea, 2010). Very hot weather, often aided by human vandals, can precipitate heath fires that annihilate large numbers of reptiles (**Fig. 3.3**). There's no doubt that 'environmental stochasticity' has substantial effects on the populations of many amphibians and reptiles.

Stochasticity
Random variation

3.5 Human regulation of amphibian and reptile population sizes

In specific cases outlined above (stopping grazing, burning heathlands) human activity has influenced essentially natural processes, sometimes with significant consequences for amphibians and reptiles at the population level. Direct human predation, however, has probably had negligible impacts on any species. Huge numbers of frogs were once collected for research and teaching purposes but the populations involved remained large at a time when many others were declining. Nevertheless, and despite caveats about detecting long-term trends reliably, there is plenty of evidence that most British amphibians and reptiles have declined substantially in recent times. It is clear that these declines were precipitated by humans but any future reversal requires identification of the specific causes.

What of the big issues in recent decades, starting with pollution and pesticides in the 1960s and 1970s, followed by acid rain and ozone depletion in the 1980s and with climate change as the current global concern? The impacts of all these threats have been studied, more with amphibians than reptiles, but there is little evidence of widespread or large effects by any of them on any species so far. Industrial pollution has damaged ponds causing amphibian declines and extinctions but on very local scales. The most widely used pesticides have little direct effect on amphibians. Acid rain has damaged ponds in the past, occasionally beyond use by common frogs and natterjacks but again mostly with local consequences, now often reversed. Increased UV radiation due to ozone depletion can kill amphibian eggs but again the damage has been patchy around the world, mostly at high elevation and nowhere demonstrated in Britain. Climate change with its trend towards milder winters has affected the breeding phenology (migration and spawning times) of some British amphibians though not as yet their population dynamics, with the possible exception of wet winters and crested newts as mentioned earlier (Griffiths, Sewell and MacCrea, 2010).

Much more important is what humans do directly to habitats. A century ago the human population of Britain was less than half its present size, farming was not intensive and few people travelled far beyond their place of birth. An early change as farming methods improved was retreat from marginal habitats such heaths and dunes where natterjacks, sand lizards and smooth snakes live. But this was not good news. Lack of grazing or other management instigated successional changes towards scrub and woodland. Together with afforestation and building developments, this reduced the habitats of our rarest species by more than eighty percent within a hundred years. The commoner species have suffered comparable problems with habitat loss and degradation since the second world war, in this case for the opposite reason. Farming started to intensify dramatically across most of the country in the 1950s. The consequent destruction of hedgerows, infilling of ponds and their eutrophication (over-enrichment) by artificial fertilisers, and widespread application of pesticides have had major impacts on all wildlife. Wildflowers, invertebrates and farmland birds as well as the commoner amphibians and reptiles have suffered major declines over the past few decades as a direct result of these changes. Urbanisation and the road network have increased concurrently. For animals of low mobility, including most amphibians and reptiles, the remaining good quality habitats are often a patchwork of 'islands' separated by intensive agriculture and busy roads, permitting little or no migration between the fragments.

Eutrophication

Over-enrichment with nutrients, especially nitrates and phosphates

3.6 Community ecology

The species assemblage and interactive dynamics in a habitat are the subject material of community ecology. Which species are regularly found together, how do they compete or contribute to food webs and how do they respond to change, such as the appearance of a new species? Ponds are particularly attractive subjects for community ecologists because they represent relatively well-defined ecosystems, and many studies that include amphibians have been published (Beebee, 1996). Sometimes species avoid competition by breeding at different times of year, sometimes they minimise predation by choosing temporary ponds, and so on. In Britain all three species of native newts can occur in the same pond, but smooth and great crested newts are found together more often than any other combination. This is probably because palmate newts do best in conditions of water chemistry (low ion content and/or high acidity) that are less well tolerated by the others. Natterjack toads avoid ponds used by common frogs or toads wherever possible because larvae of the rare species are competitively inferior. Where the common species have encroached into natterjack habitat, changes in this aspect of community structure have been so disastrous in some places that natterjacks have gone locally extinct. Understanding how communities function is therefore very important for effective conservation.

Similar considerations undoubtedly apply to reptiles though they have been little studied in Britain. All six of our native species occur on heathland, the best reptile habitat in this country; but they are not always together in the same microhabitats. Sand lizards prosper best in mature stands of deep heather, and when their numbers are high viviparous lizards are often rare or absent. Almost certainly this is because the larger lizard actively preys on the smaller one, at least in its juvenile phase. But viviparous lizards may abound close by, in shorter heather or in grass tussocks near ponds. Snakes show a similar picture. Areas of mature heath with good smooth snake populations can be low on adders, and again predation is the likely cause. Adders may be common on the periphery very close to areas where smooth snakes dominate. There is a concern that introduced wall lizards may be interfering with reptile communities in Dorset, particularly sand lizards. Once again we need to understand as much as possible about community dynamics if conservation is to work well.

3.7 Conservation

All native amphibians and reptiles in Britain are now protected by law, albeit to varying extents (see **9.1**). This sets the scene for effective conservation actions. The recognition that habitat destruction has been the major driver of amphibian and reptile population declines focuses attention on priorities for arresting or reversing these trends. Most of the surviving critically endangered habitats, notably lowland heaths and coastal sand dunes and marshes, now enjoy statutory protection as nature reserves or at least as sites of special scientific interest (SSSIs). Detailed information about the habitat requirements of all the British species together with management methods have been published by Amphibian and Reptile Conservation (ARC) as handbooks (Edgar, Foster and Baker, 2010; Baker and others, 2011). On heaths there have been efforts to reduce the extent of conifer plantations, clear scrub, kill bracken and reintroduce grazing by domestic animals. In some cases acidified sediments have been scraped out of ponds. These methods are labour-intensive and expensive but they work and have proved successful for natterjacks, sand lizards and smooth snakes. Biodiversity in heathland and dunes has benefited generally from this work with increases in woodlarks, silver-spotted blue butterflies, shoreweed and many other species. ARC has played a major role in these recovery programmes in conjunction with the statutory agencies (Natural England, Countryside Council for Wales and Scottish Natural Heritage) and many other bodies including the British Herpetological Society and those listed below (all detailed in chapter 10).

For the commoner species the scale of the problem is inevitably much greater, living as they do in a large proportion of the British landscape. A network of local efforts is needed and increasingly provided by regional Amphibian and Reptile Groups ('ARGs'), County Wildlife Trusts and charities such as Pond Conservation. Professionals and volunteers join forces to improve habitat for reptiles (**Fig. 3.4**) and create or restore high quality ponds. However, the central issue of farming policy remains. Encouraging recent developments have included wildlife enhancement schemes, especially stewardships, in which farmers are paid to improve their landholdings

Fig. 3.4 Constructing a hibernation bank for common reptiles (Lee Brady)

for wildlife as well as to produce food. Uptake has been good but may always be vulnerable to changing relative profit margins for food production (likely to increase) and conservation management (which may not). The vogue for wildlife gardens with ponds has offset losses of a few amphibian species in the broader countryside. In some cases it has proved possible to restore or recreate habitat and then translocate animals, thus starting off new populations. This may happen as part of a proactive conservation strategy (mostly for the rare species) or, less comfortably, as mitigation for commoner species when their sites are threatened with development. In this situation there is no net gain. The new site simply replaces one that will be lost, and all too often is inferior to the original.

The issue of habitat fragmentation and the need for effective conservation at the landscape level are also increasingly recognised. A patchwork of protected sites, including nature reserves, is valuable but in the long term will not be enough. It may be necessary to link together areas of high quality wildlife habitat not just to prevent early extinctions but also to provide corridors if species shift their ranges as climate change intensifies; but the practicalities look daunting. The damaging effects of roads have been popularised by 'toad patrols', volunteers taking their lives in their hands to carry migrating animals across roads on spring nights. But tunnels under roads or green bridges over them are in principle much better, because they increase connectivity among populations of many species and, unlike the patrols, function at all times of year. Substantial challenges lie ahead but at least we now know the kind of management that is needed to conserve our amphibians and reptiles effectively. There can never be too much money or too many volunteers for this task, so if you are interested in contributing, contact ARC or a local ARG (see **10.1**) for guidance on how you can help. Studies of local populations can also help by providing the information that is needed to underpin habitat management or to tackle unexpected threats such as development proposals.

4 Surveying and monitoring

One of the simplest but also most valuable ways to engage with amphibians and reptiles is to become involved with survey or monitoring programmes. As described in chapter 3, there is plenty of evidence that many of our native species have declined in recent decades but until recently there has been no serious attempt to quantify the extent of these worrying trends. Effective conservation absolutely requires this kind of information. Fortunately times are changing and we now know enough about the ecology of amphibians and reptiles to obtain useful data about populations at the local and ultimately at the national level. This will only be achieved, however, with sustained efforts around the country by as many people as possible, both amateur and professional. This chapter describes how you can contribute to this important objective and help safeguard the future of these intriguing animals.

4.1 Strategy

Surveying and monitoring are essentially similar processes. Monitoring mostly involves repeating surveys, using the same methods, at predefined time intervals with a view to detecting trends over time in distribution and population size. Any particular exercise will usually be local, involving perhaps one or a few ponds or a discrete area where reptiles might be found. Data therefore accrue for specific places but, if collected in a standard and systematic way, they can be fed into databases which over time will provide an ever more complete regional and ultimately national impression of what is going on. This should be the central objective for most people: think and act local but in such a way that the data can contribute to the big picture. Everything is now in place, in terms of both field methods and databases, to make this strategy hugely valuable as information accumulates over the years ahead.

For every surveyor some initial decisions are necessary. Will it be amphibians or reptiles, or both? Is it likely to involve species for which licences are needed (**9.1**)? And will it focus on widespread species, maybe the only type found in your area, or involve rarities (natterjacks, sand lizards or smooth snakes) which will require extra guidance and support? At this early stage a good plan is to seek advice, given your initial choice, from Amphibian and Reptile Conservation (ARC) and if possible from a local Amphibian and Reptile Group (ARG). Since 2007 a National Amphibian and Reptile Recording Scheme (NARRS), run by ARC, has been in place and this should be the context in which you plan your survey. Contact information including the relevant websites is given in Chapter 10. Many but not all counties have ARGs and if yours doesn't you can try the umbrella body (ARG-UK), which may well encourage you to set one up!

Finally at this stage comes a decision about the kind of data to collect. There are two options, not mutually exclusive. The NARRS project was set up to sample sites across Britain chosen randomly, that is, selected without reference to previous knowledge as to whether any amphibians or reptiles were present at the chosen locations. The concept, a very important one, is to look for changing patterns of local distribution (and, cumulatively, national distribution and abundance) over time without the possible bias that might arise from just selecting attractive-looking sites. On contacting ARC you will receive details of a randomly chosen site near where you live, such as, for example, 'the south-westerly most pond in a particular ordnance survey map grid square'. You will also be provided with details of what to record under the NARRS scheme, when and how often to do it, and so on. An essential aspect of this approach is that only the presence or absence of each species is required, not how many of each there might be. It can be frustrating if your 'random' site has few or no species detectable even after multiple visits, but negative occupancy data are crucial to the building of an accurate national picture, so stick with it! Just a couple of years after its instigation this methodology started to give unprecedented insights into the distribution and abundance of all our widespread species.

Another approach, initiated in the Netherlands and complementary to NARRS, involves selecting sites known to have one or more species and then making regular estimates of population sizes, not just presence or absence. In some ways this is a more satisfying procedure because there will definitely be something to find, at least in the first year. On the other hand estimating population size is inevitably more demanding than simply recording whether a species is present or not. To generate statistically robust data from which population trends can be inferred both approaches require several hundred sites across Britain as a whole, each surveyed at least once every few years. Perhaps the ideal is to combine both methods, one random site which might prove disappointing and another site with amphibians or reptiles definitely present to maintain enthusiasm. Either way, surveying is fun with the excitement of never knowing what you will find on the next visit. Combining both approaches in Britain promises to generate the best dataset on amphibians and reptiles anywhere in the world, with all the implications that brings for effective conservation.

4.2 Survey methods for amphibians

There are four well road-tested methods recommended for general amphibian surveys. These are applicable, with varying degrees of efficiency, to all of the native and most of the introduced British species. Amphibian surveys are almost always carried out in the spring months while the breeding season is under way and the animals or their offspring are in ponds or ditches.

Daylight searches

Patrolling around the edges of ponds or along ditches in daytime reveals common frog spawn as highly visible clumps or large mats, usually in just a few centimetres of water, and natterjack spawn as strings lying conspicuously on the bottom of shallow pools. Strings of common toad spawn are not so obvious but are likely to be massed together, wrapped around submerged vegetation at depths typically of 10–30 cm. Water frog spawn may be visible as small clumps among weed near the surface in relatively deep water, but can be hard to find. The presence of newt eggs is revealed when the leaves of submerged aquatic plants are bent into obviously unnatural shapes where females have wrapped them around their eggs for protection. Detection of newt eggs in this way from the pond bank is often surprisingly easy (**Fig. 4.1**) but if there is a chance that great crested newts are present a licence should be sought (see **9.1**). Daylight searching is also the simplest way of finding water frogs. They all like to bask in bright sunshine and can be seen sitting out on the banks of ponds or ditches any time

Fig. 4.1 When a female newt lays eggs, she folds a leaf over each egg (Chris Gleed-Owen)

from April through to September. These frogs jump into the water at the approach of danger (you!) and the trick is spotting them first. Binoculars are useful and detecting water frogs is very similar to locating basking reptiles.

Using a pond net

Again this is a daytime activity and pond dipping is the classical way of investigating all forms of pond life. It can be useful for finding adult newts and the larvae of all amphibians. Netting is particularly applicable where midwife toads are suspected because adults are hard to find. Their tadpoles are distinctive and unlike those of other British frogs and toads they often overwinter, becoming large by the following spring. There are downsides though. Netting is hard work and can be difficult to do effectively, especially if there is little or no submerged vegetation or so much that the net hardly moves. A mesh width of around 2 mm, not less, is desirable to minimise drag. Suppliers of pond nets and other survey equipment such as powerful torches are given in chapter 10. Speed of net movement is essential to catch newts. There is also an issue about the physical damage to weed and associated fauna that vigorous netting can cause, though this damage is short-lived. For standardisation it is necessary to net consistently. The NARRS protocol suggests a two-metre arc from a number of fixed points which should be recorded and estimated in the end as a percentage of the total pond perimeter, some of which may be inaccessible or inappropriate for netting on account of too little or too much weed.

Night searching

Walking the pond perimeter after dark probing the water with a powerful (at least half a million candlepower, rechargeable) torch is a very effective survey method. Newts tend to enter shallow, clear water at night for courtship routines. It's usually possible to identify both sexes of great crested and alpine newts and male palmate or smooth newts in a torch beam. It's virtually impossible, though, to distinguish female smooth and palmate newts without handling them. Female Italian crested newts usually have a yellow vertebral stripe, which is easy to see, but males of the two crested species cannot be separated with confidence under water. Torches are also useful for detecting amphibian larvae. Limitations of night searching include too much vegetation, especially floating plants such as duckweeds, murky water, rain and wind. All these factors limit visibility, sometimes so much that the method is useless. You can choose the right weather conditions but there's nothing to be done about the other problems.

Torching at night is also the best way to find common toads which often hide under water during daytime even at the height of the breeding season. After dark they emerge and are easily seen. The highly secretive and fully aquatic clawed frogs are also detectable at night by torchlight.

Live-trapping

For reasons that nobody understands newts of all the species present in Britain have a predisposition to enter unbaited traps set overnight in ponds. Griffiths (1985) showed how a simple, cheap and effective trap can be made by cutting off the top of a plastic drinks bottle (usually a 2-litre one) and inverting the top back into the rest of the bottle, creating a funnel similar in principle to those used for catching lobsters (**Fig. 4.2**). The trap is best set with the open end touching the pond floor, so usually near the pond edge, and the upper part either above the water surface or at least with an air reservoir so that newts caught in the bottle do not suffocate. The whole contraption is anchored with a stick, commonly a garden cane. A series of traps at about two-metre intervals, with a maximum of 25 per pond (but usually fewer), are set in the evening and inspected early the following morning. Trapping can be extraordinarily effective, regularly finding newts in ponds where all the other methods

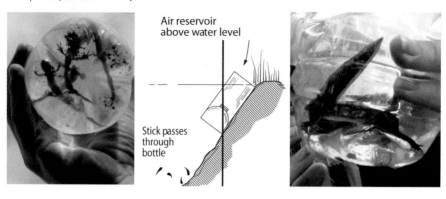

Air reservoir
above water level

Stick passes
through
bottle

Fig. 4.2 Standard newt
trap (Philip Belfield
[photographs])

fail. Clawed frogs also enter traps but if their presence
is suspected it is best to construct larger versions with
bigger funnels and with added bait, such as bits of meat.

Once again, though, there are downsides. One is
the need to visit ponds twice (evening and early the
following morning) to obtain the results. Secondly
and more important is the risk of accidental mortality
of animals caught in the traps. In warm weather when
water temperatures exceed 20°C newts sometimes suffer
even when air should be available in the reservoir,
especially if many newts accumulate in the same trap.
Usually they recover, appearing dead while actually
just being unconscious but it can take quite a while
for them to revive and they have to be protected (by
watching over them, laid out on land) during that time.
It is therefore desirable to use conventional traps only
under conditions when this problem is unlikely to arise,
so not on especially warm evenings. Even worse are the
rare events when water shrews or grass snakes enter
the traps because they invariably drown before dawn.
Happily there may now be a solution to these difficulties.
Brian Banks of Flag Ecology devised a modification in
which an 'escape window' is cut in the top of the trap. In
trials these modified traps reduced newt capture rates
but only by a maximum of twenty percent depending
on species. Unlike controls (normal traps), however, they
never caused any deaths of newts or bycatch species.
This is surely the way forward.

These are the main but not the only methods for
finding amphibians. Two more limited applications are
worth mentioning.

Calls of breeding males

In North America, with many species and lots of distinctive calls, this is a major way of surveying for frogs and toads. In Britain it is less useful. Common frogs and toads croak so quietly that they cannot be detected more than a short distance from the breeding site, and clawed frogs do not call at all. However, the choruses of male natterjacks and water frogs are both distinctive and loud and can be useful for tracking down breeding sites. Most calling happens at night but water frogs can call prolifically by day as well. Male midwife toads make their distinctive call from individual hiding places on land, not necessarily near ponds, and this is one of the best ways of establishing their presence in an area, although it's notoriously difficult to track down the individuals responsible.

Refugia

Turning over debris is not an efficient way of detecting most amphibians although many species do use such hiding places. But it is particularly helpful when seeking confirmation that midwife toads are present. These toads spend very little time in ponds (just while males visit to release tadpoles from the eggs they carry) and so adults are unlikely to be found by any of the four main methods.

Finally, a word about weather. Temperatures fluctuate wildly during the spring when amphibians are breeding but activity is generally suppressed under cold conditions. When the temperature drops to five or six degrees Celsius or below, survey methods may still reveal animals but detection efficiency is likely to be poor.

4.3 Survey methods for reptiles

Reptiles do not congregate in specific places to breed and generally do not have distinctive early life stages (except for the eggs that some species lay, but which are rarely found by surveyors). Fewer methods are therefore available for reptile survey than for amphibians and only two are commonly used. On the positive side, both methods work well from April through to September so unlike amphibian survey their application need not be confined to the spring months. Nevertheless, April and May are the best months in which to survey for reptiles because in this period they are particularly bold and conspicuous, with lots of breeding activity and lengthy basking requirements.

Daytime searching

This capitalises on the need of all our reptiles to bask in sunshine to raise their body temperatures for hunting and reproduction. Areas of habitat are inspected by walking slowly, in suitable weather and without casting a shadow in front, looking several metres ahead to spot animals soaking up the sun. Binoculars are useful aids. With practice you will identify promising spots such as small clearings exposed to the sun but in or near bushes, shrubs or long grass (**Fig. 4.3**). A good plan is to establish a survey route that takes about an hour to walk, incorporating as many likely basking places as possible (so probably not a 'direct line' transect). Care and experience combine to enable you to detect the snake or lizard before it sees you, when it will probably disappear quickly into cover. If you get just a fleeting glimpse or hear the rustle of an escapee before seeing it clearly, it is worthwhile returning at least fifteen minutes later to try again. Individuals have favourite basking sites and often come back to them, or to one close by, as soon as they feel safe.

Fig. 4.3 A typical basking spot: an open area among heather bushes (Neil Armour-Chelu)

Weather conditions are critical for high chances of success. There should be sunshine but it must not be too hot and prospects are especially favourable when sun alternates with cloud or even occasional light rain. There

is no substitute for experience both in judging the best weather conditions and for seeing the animals. For two species, slow-worms and smooth snakes, visual searching for basking animals is especially challenging. Neither is predisposed to expose itself the way our other reptiles do. Basking, when it occurs, tends to be cryptic, meaning that the animal is at least partly under cover or wound around tangled vegetation. The second survey method, described below, is far better for both of these reptiles.

Artificial refugia

This method also hinges on the need for reptiles to warm up, but by providing suitable cover it allows them to do so without exposure to predators. Flat pieces of corrugated iron ('tins') or roofing felt, at least a metre or so square, are very attractive to many reptiles when laid out in sunny positions on suitable habitat (**Fig. 4.4**). Placement of refugia requires careful consideration. Those in the middle of large open areas are less effective than those close to patches of cover, where reptiles are likely to find them more quickly. Multiple refugia are employed, ideally as an ordered array of up to 37 (Reading, 1997) but in practice usually far fewer (perhaps 12 or so) adapted according to the size and geography of

Fig. 4.4 An artificial refuge (Alan Martin)

individual sites. Refugia should be put in place at least a couple of months before the survey, giving time for reptiles to find them, and they should be taken up when the survey is complete. They can be numbered with paint and simply lifted regularly to find any animals beneath them. Use of thick gloves or a bent stick is recommended for this, otherwise fingers may by chance end up close to an adder's head. Natural refugia (such as logs, bits of plastic debris, or flat stones) are always worth checking as well wherever they occur in sunny positions.

Again weather conditions are important. It gets very hot under refugia in bright sunshine and there is little point in looking on a sweltering summer's day. On the other hand reptiles often linger longer under refugia when the sun goes in than they do when basking in the open. These artificial aids are very effective for all the snake species and for slow-worms but much less so for the limbed lizards, which are more likely to be seen on top of them than under them.

Although weather is critical for reptile survey and has been the subject of much work to identify optimal conditions, more study is still needed (see chapter 6) because the situation is so complex. In early spring, for example, reptiles emerging from hibernation bask in temperatures just a few degrees above freezing provided there is bright sunshine. Later in the year their behaviour changes, and this inconsistency occasionally makes it hard to predict good survey times, even for very experienced herpetologists.

4.4 Presence/absence surveys

The NARRS project, which started in 2007, is based on surveys that record presence or absence of each species at a location, but not the numbers of individuals. If an amphibian or reptile is found then it's clearly present but a critical aspect of this approach is deciding when a species is truly absent rather than just undetected. Fortunately there have been studies designed to address this question. Multiple sites were revisited to survey for the widespread native amphibians (common frogs and toads, smooth, palmate and great crested newts) and all the native reptiles including the rare sand lizard and smooth snake, in two consecutive seasons. From the results (for example Sewell, Beebee and Griffiths, 2010) it was possible to establish how many visits are needed through the course of a season and how many of the methods listed above need be used to end up 90% certain that if a species isn't found, it really isn't there. If a survey is to be rigorous, the conclusions of this work should be applied. Those conclusions are as follows.

Native amphibian surveys

For all five widespread species (that is, excluding natterjacks and pool frogs) there need to be four surveys over the course of the spring, each using all four of the methods listed above. The first should be timed roughly when frogs are spawning in the survey area, which can be determined by preliminary visits when the weather looks good in early spring or from other local sources of information (maybe a friend with a garden pond). The remaining three visits should be spread across the period from early April to mid May on relatively warm nights. These timings will vary across the country, collectively later as you move from south to north. An efficient way to proceed for each survey is to visit the site in the late evening to set traps, then torch after dark, return next morning to inspect traps, carry out visual survey and finally use the pond net which is applied last because of the inevitable disturbance involved. In the case of natterjacks the distribution in Britain is thought to be completely known, with the last new site discovered in 1993. However there could still be surprises. Pellet and Schmidt (2005) showed that at least four visits to a pond to listen for calling males during the main breeding season (mid April to late May) were required to establish absence with 90% certainty.

Native reptile surveys

For all six species the requirements for 90% confidence limits on absence (that is, the requirement if you are to be reasonably confident that the species is absent) were also at least four visits over the course of the season (but in this case between April and September) using both of the methods described earlier. Choice of suitable weather conditions is especially important in reptile survey and sightings usually peak in April and May.

Non-native species

Survey requirements have not been investigated as thoroughly for these amphibians and reptiles as for the native species, except for midwife toads where again four visits to listen for calling males in spring were sufficient to establish absence with 90% confidence (Pellet and Schmidt, 2005). All the other species are usually reported incidentally when applying the survey protocols outlined above for native species.

Four visits seems like a magic number but in all these survey strategies it is of course not necessary to carry out so many trips if all the species likely to be present (say the four widespread reptiles on a sunny downland slope) are found within the first couple of surveys. There are then no 'absences' left to prove. Standard forms are available from ARC on which to report data collected for NARRS, which can also be done on line. They include suggestions about reporting additional information from the survey sites, all of which should prove useful for later scientific analyses. ARC will also help confirm (or otherwise) uncertain identifications. Verification of records is very important for the rare species (natterjacks, native pool frogs, sand lizards and smooth snakes) and all of the non-native species. Wherever possible it is best to obtain digital photographs to submit with the records.

4.5 Population size surveys

Going beyond presence/absence and assessing the sizes of amphibian and reptile populations is inevitably more demanding of time and effort. In most cases it is impossible to obtain accurate total census estimates, short of a detailed scientific study not feasible for most volunteers. This is because it is nearly always impossible to detect every individual in a population by observation alone. The percentage found will almost invariably be less than a hundred, often much less, and will vary among species, sites and seasons. Nevertheless it is feasible to make relative estimates of population sizes, making comparisons between sites but especially between years, and these can be very valuable for detecting trends over time. The assumption here is that detection rates (though not known) will be reasonably constant. This is usually safer for comparisons of the same site over time than for comparisons of different sites, but even this assumption is certainly not valid all the time. In some springs, for example, not all toads (common or natterjack) will breed if the weather conditions are unfavourable. In most cases, however, the real trend is strong enough in relation to the annual fluctuations (on which it is superimposed) to produce credible results. This method has been employed successfully in the Netherlands since the late 1990s, at several hundred locations, to assess population changes at the national level in all that country's amphibians and reptiles (see www.ravon.nl).

The estimation of population size often requires the application of species-specific methods. Because of their distinctive life cycles it is not possible to generalise across, for example, all the amphibians. The most reliable approaches are listed below together with the species they apply to. Whichever method is employed it is normal (and good) practice to visit the site at least three times during the breeding/main activity season and use the highest count as the final estimate for the year.

Spawn counts

This is very suitable for common frogs and natterjacks. Frogs congregate for spawning and simply counting the clumps gives an estimate of female population size. This is easiest when the eggs are fresh laid. After a while the clumps merge into one another but it is still possible to calculate the likely original number based on the area of the spawn mat (**Fig. 4.5**). Natterjacks usually lay their spawn separately from each other, making counting straightforward. However it is necessary to visit the ponds regularly (at least once every ten days) through the six- to eight-week breeding season to obtain a cumulative count and thus an estimate of female numbers. This procedure comes closer to yielding accurate population size estimates than methods available for any of the other species because spawn detection rates are usually high.

Head counting by torch at night

This is appropriate for common toads, which tangle their spawn together in uncountable masses. Toads breed explosively and it is essential to visit the breeding site several times over the breeding period (usually a couple of weeks), moving steadily along the bank and recording every individual. Many animals become active at the water surface after dark and can be counted even in ponds that are too weedy or too murky for newts to be detected efficiently. Clicker-counters are invaluable for this. The highest count of the season is taken as the relative estimate. Be careful to choose mild nights with little wind, conditions under which the highest proportion of the assembled animals will be visible. Independent studies suggest that even on the best nights fewer than 50% of the toads actually present will be recorded.

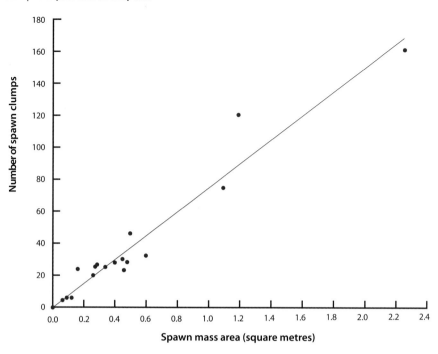

Fig. 4.5 Relationship between number of common frog spawn clumps and area of spawn mat. © JNCC. From: R.A. Griffiths & L.D. Brady (1996) Evaluation of standard methods for surveying common frogs (*Rana temporaria*) and newts (*Triturus* spp.) JNCC report 259, 1996.

Torch counting for newts

All the newt species are easier to detect by night than by day in ponds. As with toads, an observer moving slowly round the pond and registering each individual gets some indication of population size. This method works best for the distinctively large and dark great crested newts. Unfortunately alpine newts are readily confused with them at night, because although they are smaller than crested newts, they also look very dark in torchlight; so extra care is needed in the unlikely event that this non-native species is present. The two small species are more problematic. A 'total small newt' count is fine if you also know (maybe from trapping) that only smooth or palmate newts are present in the pond. If both occur together it is usually possible to distinguish the males but not the females. Inferences therefore have to be made from the relative numbers of males.

Trap counts of newts

Many ponds are unsuitable for torch-counting newts because of high opacity or too much submerged vegetation. Traps therefore have the advantage of being more broadly applicable than torch counting and there are very few ponds where trapping is not possible. Other advantages are that species can be identified with certainty and that consistency between years is likely to be better for traps than for torch searching, which is vulnerable to trends of changing plant cover or turbidity. On the other hand, in clear ponds counts obtained by torch searching can be much higher than numbers caught in an array of traps. For both torch counting and trapping, normal practice is to record the highest number seen or caught on one occasion after carrying out the assessment several times through the newt breeding season (April to mid May in most places). A sensible strategy for newts is to use both torch searching and trapping wherever possible and compare the results (see also chapter 5 for suggestions on further study). In general we have little idea about how torch or trap counts relate to actual newt population sizes in the ponds because the relationship must be hugely variable. Indeed it would be useful to determine how the two methods compare with respect to number estimations in a particular pond. Which is more consistent (shows least variance) over multiple visits? Are any differences between the methods consistent between species?

Head counts
This method is applicable to all the reptiles. Whether walking transects to look for basking animals or turning over refugia to find them, it is the cumulative number of each species that goes on the record. Once again it is standard practice to use the highest number seen in a series of visits, preferably during the optimum months for high visibility (usually April, extending into May) and of course timing the survey visits to take place in suitable weather as far as possible. Head counts of reptiles are even more difficult to relate to true population sizes than is the case with most of the amphibian methods. A few detailed studies, for example with sand lizards, have indicated that under optimal conditions perhaps half the population known to be present from capture-mark-recapture (CMR) studies may be out basking at the same time. But optimal conditions are rare and the proportion visible will usually be much smaller. Furthermore neither visual searching nor trapping is likely to include all the places within a site where reptiles occur.

4.6 Using the data

There are several reasons for carrying out amphibian and reptile surveys, not least of which is the simple pleasure of finding out what's in your local pond or on that wild hillside. Usually other interesting animals and plants turn up along the way, all adding to the fun. At another extreme, commercial consultants regularly survey sites, especially those where any of the fully or partially protected species live. Potential developers have a legal obligation to find out what's there and if necessary to arrange mitigation (usually collecting and moving the animals somewhere else) before any construction work can begin. These surveys have not always been carried out with the rigour necessary to prove absence, perhaps (a cynic might argue) because developers would prefer that a rare species is not found, with its inherent implications for the cost of mitigation or even the possible failure of a planning application.

For amateurs and many professionals, though, a major reason for carrying out survey work is so that the information can be used to inform future conservation. The survey findings obviously cannot be used in this way unless they are made widely available. There are several ways to do this. Firstly, and preferably, the results can be submitted to ARC under the NARRS scheme. They will then contribute to the national effort to monitor all our amphibians and reptiles over coming decades. ARC will also forward them to the national database for a wide range of taxonomic groups in Britain, the National Biodiversity Network (NBN) gateway (http://data.nbn.org.uk). You cannot submit your results directly to the NBN; only registered members (generally organisations) can do that. ARG-UK also has a recording scheme to which survey results can be submitted (see the ARG website, **10.1**). Then there are, in many parts of Britain, recording schemes run by local authorities and/or the wildlife trusts, and these too will be pleased to receive survey data. You have the right to decide who can see or use the information you provide, and you will often be asked to agree that the data can be made widely available. Some recorders perceive a dilemma here: why should commercial consultants have access to information, provided by you free of charge, for work they are being paid for? Policies vary among databases but some do charge for access to the information they hold. On the other hand there is a big advantage to having all possible data available for consultation by planning authorities. Your contribution might help to save a site from destruction by developers.

Once your data have been submitted, especially if they go via ARC, you will have the satisfaction of knowing they will contribute to future conservation strategies. For either presence/absence records or population size estimates, a large network of sites (more than four hundred of each type in Britain) is needed for statistically sound detection of national status changes in time frames short enough to allow conservationists to respond and take action if it becomes necessary. So surveying will turn into monitoring; you or your successors will add data as the years go by. **Fig. 4.6** shows an example of how this has already produced surprising revelations in the Netherlands. Who would have expected to find sand lizards increasing but viviparous lizards declining?

Fig. 4.6 Changes in population size estimates of lizards in the Netherlands from 1994–2007, derived from volunteer survey data coordinated by Annie Zuiderwijk and Ingo Janssen (2008) of RAVON.

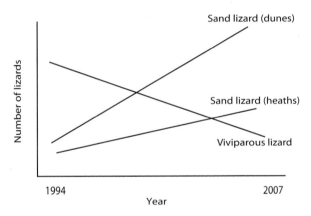

5 Studying amphibians

A great attraction of amphibians and reptiles is the ease of studying them both in captivity and in the field and finding out new things about their lives. Arguably the simplest type of investigation, surveying for presence and abundance, is described in detail in chapter 4, but there are plenty of other interesting avenues open to amateur investigation. It's surprising but true that despite a major effort to understand the ecology of all the British species over the last few decades, there is still a lot to learn and much of it could be directly applicable to conservation. Garden ponds are extraordinarily useful in this context because many investigations described later in this chapter are well suited to amphibians breeding in them. It's incredibly convenient to have an essentially 'wild' population just outside the back door. This chapter focuses on possible research topics with amphibians, while chapter 6 covers comparable ground for reptiles, and chapter 7 raises some questions best tackled by a group of people, such as a school or a natural history society, with both amphibians and reptiles. A specific aspect of amphibian biology, the complex life cycle, enriches research options considerably because each stage, from egg to adult, is amenable to investigation in various different ways. Most of the ideas in this and the following chapters are exactly that, ideas, open to different interpretations and applications (for example among several species). No doubt the imaginative enthusiast will think of others. It is often valuable for the same kind of study to be repeated in many places to obtain an overview. How does life in Scotland compare with that in southern England? Many aspects of amphibian and reptile ecology are strongly influenced by local climate and geology. Sometimes, given the relatively low diversity of amphibians and reptiles domiciled in the UK, I have raised questions targeted at particular species where these could provide useful insights.

Much of the work outlined below can be done very cheaply, but there is a scale of expenditure and effort depending on exactly what you choose to do. Some items such as digital cameras and garden ponds are not considered because most people will already have purchased or installed these for more general purposes. At the low end of commitment, bottle traps (see **4.2**) can be produced quickly and easily with minimal expenditure (and the drinks they contain are a bonus!). Travel costs to local study sites should also not prove extravagant. A little more expensive are strong, durable pond nets, high power torches and containment facilities such as small tanks for studying various life stages in captivity (**Fig. 5.1**). Small plastic aquaria are ideal for this purpose and can be purchased cheaply. At the top end but still feasible for the dedicated enthusiast are microscopes, PIT tags and data loggers, all of which can add up to several hundred pounds, but it may be possible to borrow microscopes from a local school or university. A few of the suggestions in this and the next chapters are unashamedly ambitious, approaching the kind of work carried out by professional scientists, but they are nevertheless feasible especially where it is possible to work as a group.

5.1 Life in the pond: from egg to metamorph

The basic biology of amphibian development has been described in textbooks for over a century. Because it is so easy to watch the transition from fertilised egg through development of embyos, larval development and ultimately metamorphosis into a young frog or newt, the whole process has long been popular with researchers in fields as diverse as molecular biology, developmental biology, physiology and evolution. Yet only recently have some ecological aspects of this fascinating process come to light. It was surprising to discover, for example, that tadpoles of many species change not only their behaviour but even their physical shape in response to the presence of predators (McDiarmid and Altig, 1999). Amphibian development is easy to study, and its ecological implications and relevance to conservation make it an excellent subject for amateur investigations. There is still great scope for new discoveries.

Fig. 5.1 An array of small plastic aquaria suitable for experiments at home, each about 40 x 20 x 20 cm, with sand or soil substrates and submerged plants, set up in an outhouse (near a window) or outside with a mesh covering to exclude predators

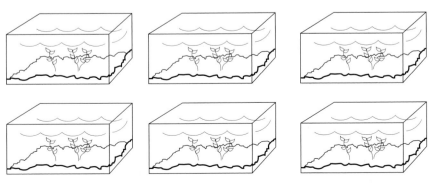

Factors affecting spawn mortality

Frog and toad spawn usually survives well over the period between deposition and hatch, typically a couple of weeks or so. Usually, but not always. Sometimes spawn mortality is obvious. More often it only becomes apparent after careful study. In either case an investigation of its causes can be valuable. Spawn can't be palatable to many predators because despite being conspicuous it is usually left alone. Yet sometimes it does suddenly disappear entirely; either it dies off with the embryos turning white and decomposing within the jelly or, bafflingly, just the embryos disappear leaving the jelly untouched. People with garden ponds regularly report these events, for which there is often no convincing explanation. What is going on? Various hypotheses could be tested. Perhaps some clandestine predator is at work, maybe there is a pathogenic microorganism out there, possibly the water quality is affected by pollution, or the pond may be relatively cold (shaded from the sun), making spawn vulnerable to sudden temperature drops, or even ice formation, in spring. If you know of a pond where this type of early mortality has happened recently (ideally within the last year), some spawn could be removed immediately after laying and its fate in aquaria compared with that of spawn left in the pond. The number of embryos that die or disappear should be recorded daily.

Data loggers can be employed to record water temperatures continuously for subsequent download onto a computer, or regular readings can be taken with a normal thermometer. Check the pond at night to look for nocturnal predation by newts, fish or flatworms creeping into the spawn jelly. Watch out for the damaging effects of ice on embryos exposed to it and record the presence and extent of fungus (*Saprolegnia*) infection if it occurs. This pathogen is common and produces white, threadlike hyphae that spread across or around amphibian eggs. Is its abundance related to water temperature or other factors? Aquarium conditions can be varied in experiments to investigate effects of water quality and temperature. Spawn can be reared in pondwater taken from where it was laid, water from a different pond, tapwater (allowed to stand for 24 hours before use to remove any chlorine present), in the open next to the pond (so experiencing similar environmental conditions) or indoors, and so on. Each condition should be replicated at least three times to generate reliable results.

There's no need to confine this approach solely to ponds where unexplained spawn mortality has been noticed. Monitoring spawn from ponds with no history of problems will produce useful information about the rates and causes of mortality in apparently typical environments. Is there much variation in spawn survival among ponds in your area? If so, why? What happens if you move spawn to a different part of the pond from where it was laid (for example to a shadier location or into deeper water) or even to an entirely different pond that the amphibian didn't use? Again, always think of replication. Try to find several unused ponds for the latter experiment. Both common frog and common toad spawn is amenable to these types of investigation and results will be valuable for future conservation if, for example, dangerous situations are identified that could be ameliorated in garden ponds. Be prepared for some results to be inexplicable, however carefully the work is planned. Virus infections that kill eggs won't be identifiable by amateur study. You may end up simply ruling out most potential causes of mortality, but arguably this is all the more interesting if it implies that some previously unknown pathogens are out there demanding professional attention.

Although these kinds of experiment are simplest with frog and toad spawn, studies are also possible with newt eggs, albeit with some extra difficulties. Unlike the situation with frog and toad spawn, predation is probably a major cause of mortality for newt eggs in many ponds, and survival rates are likely to be more variable. Newt eggs are hard to observe individually over time in natural ponds and it can be challenging to collect enough for experiments. Identification is a potentially big problem because eggs of the smaller newts (smooth, palmate and alpine) are indistinguishable (see **8.2**). Ways round the last problem include studying ponds where you know only one of these species is present, or catching females and then allowing them to spawn in individual aquaria.

Despite these problems it is possible to carry out interesting research on newt egg survival both in ponds and in aquaria. In ponds individual eggs, or groups of eggs on a particular plant, can be enclosed within fine (1 mm) mesh cages that exclude most potential predators.

Survival to hatch can then be compared by regular inspection with a set of unprotected eggs on plants marked for future reference with (say) a garden cane. Using a selection of ponds you can compare survival among sites and between species and relate this to types and abundances of predators you also find in the ponds. Aim to detect any fish present, frog or toad tadpoles and predatory invertebrates (quantifiable by netting) such as dragonfly nymphs and diving beetle adults and larvae. As with frogs and toads there is also the option of moving eggs, attached to leaves on intact plants, to ponds not containing newts to assess comparative survival rates. In aquaria, experiments like those suggested for frog and toad spawn can be designed to investigate mortality in the absence of predators. If you are starting off with captive animals in the tanks remember to remove the newts when they have laid their eggs, because adults regularly consume their own offspring. Remember, too, that if you are working with great crested newt eggs (for which you will need a licence, as described in **9.1**) 50% are bound to die due to the bizarre nature of their chromosome arrangements (see **2.1**). Any extra mortality will inevitably be superimposed on this background level.

Chromosome

A structure in the cell nucleus carrying a set of genes

More information about factors influencing the survival of newt eggs would be helpful for practical conservation. In aquaria newts could be offered different species of plants to lay their eggs on. Do some support higher survival rates than others? The answers to such questions could inform managers creating ponds to support newt populations.

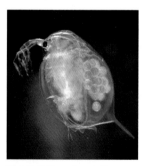

Fig. 5.2 Female adult water flea *Daphnia magna* (Duncan Hull)

Working with tadpoles

The larval stage is a period of high mortality for all British amphibians, and larval survival is therefore an important determinant of population size and viability. The more we know about it the better we will understand population dynamics. Larvae of frogs and toads are very easy to keep and study in captivity because they will feed on a wide range of readily accessible and cheap materials, such as fish flakes, boiled lettuce leaves or pellets sold for pet rodents. Newt larvae are more difficult because they require live food such as Daphnia (**Fig. 5.2**, from any respectable local aquarist shop!) or other small pond invertebrates.

Although much work has been published on amphibian larval growth and survival, this is one of those situations where more is always better. In any case, some important questions remain to be addressed. It's well known that fish can be major predators of amphibians and a conservation issue hinges around stocking ponds with fish for angling purposes. But there are many unanswered questions. Are all fish species equally problematic? And how do amphibians rank in their vulnerabilities to various fish species? What ameliorating factors may allow amphibians to persist in fishponds? Do other pond features such as extent of aquatic vegetation growth have big effects on tadpole survival rates? And why do entire cohorts of tadpoles sometimes disappear for no apparent reason?

Cohort

A group of individuals born in the same year

Predation trials with vertebrates are illegal when deliberate exposure is involved, but in ponds it is possible to glean relevant information without breaking the law. A critical issue is how to estimate tadpole survival in the wild. With frogs and toads it is usually feasible to estimate how many eggs are laid and how many survive to hatch because spawn is accessible and (with patience!) eggs can be counted, including those that die or are otherwise lost during development. The simplest way to do this is to count the eggs in a few small samples of spawn jelly or string, measure the volume of the samples, then multiply the number of eggs per unit volume by the estimated volume of spawn in the pond (easier with frogs than toads in most cases). This gives the starting estimate but after hatch, when the tadpoles disperse, assessment of survival becomes more difficult. Common frog and toad tadpoles are readily attracted to pieces of meat suspended in ponds, and one approach worth trying is to provide this food at a number of stations around the pond at regular intervals through the spring. After 15–30 minutes tadpoles accumulate around the bait and can be caught and counted. Of course this will only allow estimates of relative changes, as just a fraction of the tadpoles will turn up. Nevertheless data on survivorship will accrue as development progresses. In some ponds without much weed it may be possible to make relative estimates just by netting (especially for toads). Both these method, baiting and netting, also provide interesting extra data showing which parts of the pond are favoured by tadpoles at various times through the spring months. By measuring samples of the tadpoles

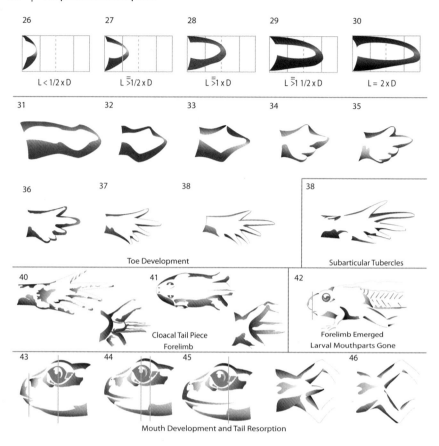

26   27   28   29   30

L < 1/2 x D   L ≥1/2 x D   L ≥1 x D   L ≥1 1/2 x D   L = 2 x D

31   32   33   34   35

36   37   38   38

Toe Development   Subarticular Tubercles

40   41   42

Cloacal Tail Piece   Forelimb Emerged
Forelimb   Larval Mouthparts Gone

43   44   45   46

Mouth Development and Tail Resorption

Fig. 5.3 Developmental stages of frog and toad larvae (Gosner, 1960). With permission from *Herpetologica*, Allen Press Publishing Services

you will also get an idea of their growth rates and rates of development. These are not necessarily the same because growth and development do not always correlate. Development has been separated into more than forty 'standardised' stages, mostly based on limb development (**Fig. 5.3**). Another worthwhile approach is to construct wood and plastic-mesh cages (**Fig. 5.4**), immerse them in ponds with their bases dug into the substrate early in spring and add tadpoles soon after hatch. Within the confines of cages absolute numbers of tadpoles can be monitored throughout development. Various treatments are possible such as removing any potential predators at the start, adding weed or extra food to some and not to others, exposing them to different areas of shade and so on. Relative survival can also be compared with that of tadpoles living freely in the pond outside the cages.

In most cases the abundance of larvae will decline

Fig. 5.4 An experimental cage partly immersed in the shallow margin of a study pond, including a removable lid to exclude predators

rapidly through the spring as mortality factors take their toll. Eventually numbers will be too small for reliable estimation outside cages but it is desirable to assess whether some survive to metamorphosis. Freshly emerged froglets and toadlets are secretive and unless they emerge in large numbers (which sometimes happens) they can be hard to spot. A useful trick is to place small refugia such as roof tiles around and close to the pond edge, with access underneath for the tiny animals to hide under. Then inspect them regularly for a couple of weeks when you anticipate that froglets or toadlets are leaving the water, commonly around midsummer.

Using these methods it is possible to assess the fate of frog or toad tadpoles in ponds with or without particular species of fish, or differing with respect to any other variables (such as weed type and cover) that you care to choose. Comparable studies with newt larvae are more difficult. Newt eggs are normally impossible to count and both eggs and larvae of the small species (smooths and palmates) are impossible to distinguish. Nor are newt larvae attracted to bait. But information about their survival is important for conservation and not completely impossible to obtain. Systematic netting of aquatic vegetation at intervals through the spring and summer can provide data on relative numbers, if you accept that smooth and palmates will have to be considered collectively unless you know that the pond is used by only one of these species. Special care is needed when great crested newts are present because their larvae are easily damaged by rough netting and you will of course need a licence. Again you can put down refugia around the pond and inspect them, in this case from late July through to September to assess survival to metamorphosis. All the methods suggested above for free-living tadpoles in ponds are semi-quantitative, meaning that absolute numbers will not be determined and nor will the range of error. Even so, semi-quantitative data are better than none, and can provide valuable insights into what goes on in a pond ecosystem.

Aquarium-based studies with replicated sets of conditions can provide useful extra information, particularly with frog and toad tadpoles, which are easy to rear in captivity. To be legal this kind of experiment should not deliberately generate stress, for example by adding predators or restricting food supply. However, growth rates and survival can be compared using different types of food, water from different sources, various water depths, varying densities of aquatic plant species and so on. Aquarium conditions can never completely mimic wild habitats, but the results will generate insights about what might be happening in the ponds. What these experiments gain in scientific rigour has to be balanced against an inevitable reduction in realism. Their main advantage is the ability to control and replicate conditions far more effectively than is possible in ponds and thus they allow potentially important factors to be investigated individually.

Aquarium-based studies with newt larvae and fish could add to our understanding of which species of fish are the most dangerous. Behavioural investigations are possible with newt larvae separated from the fish by a mesh screen. Newts detect the presence of predators by sight and smell, both of which still operate with a screen in place. Most of the experiments demonstrating responses to predators have involved predatory invertebrates such as dragonfly nymphs (McDiamid and Altig, 1999). Behavioural changes include reduced overall activity and greater propensity to hide under cover. Over a longer period (weeks) body shape can alter to a type better suited to cope with the type of predator present. Do larvae show comparable effects in the presence of various species of fish? Maybe they even change shape. In response to fish this might involve streamlining or increased tail muscularity to facilitate rapid escape. All of these potential changes are readily measurable in aquaria.

Fig. 5.5 Smooth newt
Lissotriton vulgaris
metamorph (Philip Belfield)

Metamorph
An amphibian immediately
after metamorphosis

A specific question highly relevant to conservation concerns water frog larvae. These are notoriously difficult to monitor. Soon after hatch they become very secretive, are hard to find by netting and rarely visit the meat baits used for common frog and toad tadpoles. Where do they go and what do they eat? Because of this problem it has proved very difficult to monitor breeding success of the native pool frog population accurately. As far as we know all water frog tadpoles (including those of introduced pool, marsh and edible frogs) are similar in this respect. If you live close to any introduced water frogs habitats, aquarium trials could test a wide range of possible foods, plant and animal, to find whether any might be useful as a bait to monitor the native pool frogs. This experiment would be simple but potentially very valuable.

5.2 Juvenile life

The period between emerging from the pond and returning to breed two or three years later remains the least understood part of the amphibian life cycle. The animals are too tiny for attachment of any tracking device and too small to find easily, and they often disperse rapidly over large distances. Mortality rates are very high, especially in the first year of life, and probably comparable with larval death rates. Many terrestrial invertebrates such as spiders and ground beetles eat metamorphs, and desiccation must also be a significant risk in summer for tiny amphibians with such a large surface area to volume ratio. These large and undoubtedly variable mortality rates are likely to be major drivers of population dynamics and thus of population viability. Furthermore these juveniles often disperse over long distances and colonise new ponds. This dispersal is vital to maintain genetic diversity and to ensure long-term population survival. How can amateur study add to the sparse fund of knowledge in this important area?

Individual identification

One technical but potentially revolutionary contribution would be the evaluation of high resolution digital photography as a tool for tracking even the tiniest amphibians after metamorphosis. In at least some species there are individual morphologies, essentially 'personal footprints' that may remain recognisable throughout life.

These could be wart patterns for common and natterjack toads; upper body colour patterns for common and water frogs; or belly spot patterns for great crested newts. We already know that these belly patterns change during growth in crested newts. How often can individuals still be recognised a year later, or as adults? One way to find out is to rear animals in captivity, photographing likely useful areas of the body at regular intervals. Captive rearing is practicable for all our species but requires some effort. The animals need housing in a suitable terrarium with damp soil or sand, some vegetation, a cover to prevent escape and rather a lot of food. They can be given hatchling crickets (available from commercial suppliers, 10.1) when they are very small, progressing to larger prey as the amphibians grow. They can be hibernated (in a box in the fridge) or kept awake at room temperature with continuous feeding. It will be very valuable for future field studies if individual recognition of metamorphs proves reliable. The ability to identify individuals reliably should help us to address important questions such as how many return to breed in their pond of birth and how many disperse more widely and colonise new ponds. There is very little information on this subject because it is so difficult to follow individuals from metamorph to adulthood, but it is of potentially high importance in conservation planning.

Dispersal

How might metamorph dispersal be tracked? Is movement away from the pond directional or random? Are particular microhabitats preferred? Artificial refugia of different kinds (such as roof tiles, pieces of tin, roofing felt) could be tested for their attractiveness to juveniles, initially by laying them around a pond edge before metamorphosis begins. This would also indicate whether emergence is random around the pond or concentrated around particular sections of shoreline. Then the preferred type of refugia can be distributed at varying distances from the pond and in different microhabitats; perhaps in grass, on bare soil, in shrubberies or in woodland. Initial investigations could be carried out around garden ponds, but dispersal will almost certainly advance rapidly beyond the confines of one garden. Ponds in the wider countryside offer prospects for following juveniles further but there is a law of diminishing returns, especially if dispersal

Fig. 5.6 Fence and pitfall trap system

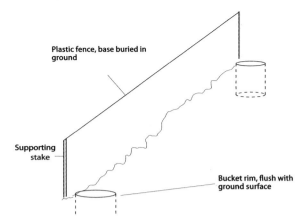

Plastic fence, base buried in ground

Supporting stake

Bucket rim, flush with ground surface

directions are random. In such cases the numbers of animals per metre of transect (at right angles to the direction of movement) will fall off as a function of the square of the distance from the pond, even in the absence of any mortality. High death rates in the early stages will also greatly reduce the chances of detection as the small animals move away. However, only trials will tell whether this kind of study is practicable. If dispersal direction is not random, the prospects of finding concentrated movements along specific trajectories are improved. Metamorphs of at least some species (specifically great crested newts) certainly do sometimes head in particular directions towards especially favoured habitats (Malmgren, 2002).

Another way to find emigrants at increasing distances from the pond of their birth is to use low barrier fences a few metres long, made of plastic sheeting and dug just slightly into the ground (**Fig. 5.6**). These obstruct movement and force the animals to move along trying to find a way round. Pitfall traps buried up to the rim at each end then catch them. These traps can be plastic buckets with tiny holes in the base (so that they don't flood if it rains) and with bits of tree bark or other cover on the bottom under which the juveniles can hide. Small wooden ladders that any small mammals that fall in can use for quick escape are also a good idea. The traps should be inspected early every morning for as long as they are in place. Obviously the longer the fence, the greater the chance of interception. This simple apparatus can be replicated at different points around the pond and at various distances away from it.

5.3 Studying adult amphibians

Frog and toad breeding behaviour

The assembly of adult amphibians in ponds during the spring breeding season makes them wonderfully accessible for observation and study. Not surprisingly, therefore, there has been a huge research input over the years and much has been published on the behaviour and reproductive ecology of all the species found in Britain. But this doesn't mean there's nothing left to discover. In some cases it is valuable to have 'more of the same' – is a behaviour seen in one part of an animal's range the same in other places? But there are also experiments that have not yet been done (or at least not reported). A few suggestions follow.

An intriguing but controversial aspect of frog and toad breeding behaviour is the relative importance of male competition and female choice in mate selection. In the 1970s a study with common toads gave evidence for assortative mating (Davies and Halliday, 1977). When the lengths of male and female toads paired in amplexus were compared during spawning, there was size matching – large males were with large females and small males with small females. Size matching makes sense for optimising fertilisation efficiency because the male and female cloacas would then be close together during egg and sperm emission. Male toads wrestle for females, and if male competition dominated breeding activities the expectation was that large males should always win, whatever the size of the female. So how could this situation arise, with small males sometimes winning? Davies and Halliday (1977) offer some ideas about how this might happen but subsequent studies on other populations produced a range of different results. In some large males always did dominate while in others pair sizes during spawning seemed completely random.

One way of resolving these differences is to obtain data from a much larger number of populations than have been studied so far. Garden ponds offer a superb opportunity for taking this forward. Their amphibian populations are relatively small, making it possible to track the individuals that turn up. And it's easy to keep a very close watch, just going outside the back door every night. Although assortative mating has mostly been described in common toads, there are reports of the same phenomenon in common frogs too, so even if a

Assortative mating
The pairing of males and females with similar characteristics (in this case, length)

garden only has frogs (as many do) a worthwhile investigation is still possible. Measuring animals either individually or in pairs is simple enough. Males hang on very tightly and will not usually let go while you measure them and the females beneath (use a ruler and take the snout-cloaca distance, in millimetres). By monitoring the pond continuously it is possible to determine whether the larger males turn up first and whether or not assortative mating is the eventual outcome when spawning starts. If it is, how does it happen? Watching frogs or toads after dark allows you to see directly how males try to displace each other. Mating frogs and toads are relatively resilient to disturbance, but using red light from a modified torch helps to avoid interfering with their behaviour. Experiments are also possible. Gently separate some pairs, put individual females in buckets with water, add pairs of males of differing sizes and monitor what happens. One explanation suggested for the differences between previous studies is that weather can affect the outcome. Mild springs may allow time for pairs to sort themselves out and match sizes before spawning, whereas if a cold spell delays the start of breeding everything happens very quickly when the temperature rises, and pairing is perhaps more random. Studies of mating behaviour at multiple sites over many seasons with varying weather conditions should eventually provide a consensus view of how sexual selection works in frogs and toads.

Another intriguing aspect of behaviour at this time concerns the choice of spawning site, which is determined in all species by the female. Once the first clump of common frog spawn is laid, other females congregate at the same place to deposit their eggs. Common toads do much the same with their spawn strings. If that first clump is moved, very often the other females will still go to it for spawning. It's as if the choice of the first female is automatically accepted by the rest as a suitable spot. But what are the limits of this behaviour? The tolerance of female frogs could be tested by moving the first clumps to a range of sites, perhaps differing in depth or shade cover, or even to a neighbouring pond known to be frogless. This can be followed up by putting some spawn back at the originally chosen site and comparing percentage survival of the eggs to hatch. Was that first female making the best decision?

Fig. 5.7 Plastic strips attached to a stick, to attract newts for spawning

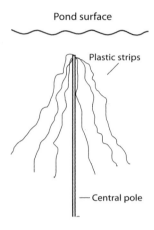

Pond surface

Plastic strips

Central pole

Newt breeding behaviour

Experiments of the type described above for frogs and toads are not possible with newts because these lack the amplexus embrace, lay eggs one by one and have internal fertilisation. However, there are alternatives. Newts readily spawn on artificial constructs (**Fig. 5.7**) in which strips of plastic (anything plastic will do, but preferably opaque rather than clear and with each strip about 5 mm wide) are attached to a central pole and immersed in a pond. Placing a set of these at various sites in a pond should show whether there are preferred spawning areas and if so, what features (such as the type or extent of vegetation, or water depth) they correlate with. It would also be useful to monitor survival of eggs to hatch on these artificial substrates, either where they were set or after they have been moved to another part of the pond. Once again, this would make it possible to assess how well females choose their egg-laying sites. A better understanding of optimal spawning sites for frogs, toads and newts could have applications for conservation managers in the design and creation of new ponds.

Fish introductions, accidental and deliberate, increase every year and pose a significant threat to great crested newts, our most vulnerable widespread amphibian. Predator trials are out of the question, but it is important to know whether some fish are more dangerous than others. How do newts respond to the presence of fish? There is some evidence that newts can detect fish in a pond and then avoid entering it to breed. But which fish elicit this response? Remember that a licence from the statutory agency (Natural England, Countryside Council for Wales or Scottish National Heritage) is necessary for any work on great crested newts, but informative studies can be performed with any newt species.

Adult female newts, mated and ready to lay eggs, can be placed in an aquarium together with various species of fish and given a platform so that they can choose to leave the water (**Fig. 5.8**). It is necessary to provide refugia (such as bits of tree bark) on the platform for them to hide under, and to cover the aquarium with a mesh lid to prevent escapes. Adult newts will not be at risk provided the fish are small specimens! It is the eggs and larvae that suffer predation. The rate at which newts leave the water, and whether and when they lay eggs, can be compared between these fishy aquaria and fish-free controls. Eggs should of course be removed immediately after laying, or the fish will eat them. A useful range of fish species would include aggressive predators such as perch (*Perca fluviatilis*), small ones like three-spined sticklebacks (*Gasterosteus aculeatus*), and omnivores such as goldfish (*Carassius auratus*), crucian carp (*Carassius carassius*), Koi and common carp (both *Cyprinus carpo*) and tench (*Tinca tinca*). It would be very useful to know which, if any, are perceived as safe by the newts.

Sexual selection in newts has also been widely studied, mostly based on observations in captivity of female choice when males of different 'quality' (such as size or height of crest) attempt to court them. Because there is no physical embrace it is less straightforward to determine the male 'winners' in these experiments. Careful and persistent observation is necessary to find out which male is accepted, as judged by the female going through a full courtship routine and taking up his spermatophore. Females caught in a pond in spring are very likely to have mated already and although multiple paternity is common it may take some time before they accept another male. Nevertheless, interesting research is feasible. One can start off with virgin females that have not yet encountered any males by using a fence and pitfall trap system, as shown in **Fig. 5.6** but with the traps on the side furthest from the pond. The whole setup must be erected before the breeding season begins, in February or early March. Females falling into the traps will be migrating to the pond and as yet unmated. In aquaria they are likely to respond quickly to male advances, but even without this ideal situation (that is, if you just catch newts already in a pond) experiments can still work; they will probably just take longer. From this starting point various hypotheses are testable. Does the number of available males influence how quickly a female mates? Does she choose according to some feature such as male body size? Although experiments along these lines have been reported before (Gabor and Halliday, 1997), there have been few attempts to follow through and investigate how many eggs are subsequently laid. It may be easier to measure this than to watch

Fig. 5.8 Aquarium with escape platform so that newts can leave the water

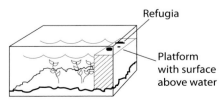

Refugia

Platform with surface above water

patiently to confirm courtship success, and egg number is in any case a more critical indicator, revealing the final outcome of breeding behaviour. Does the quality or number of males influence the number of eggs eventually laid, the timing of oviposition, or egg viability as judged by hatch rate? Eggs should be moved to another tank as soon as possible after they are laid, so that their parents don't eat them! The experiment will take longer than one based on courtship observations, since egg-laying can continue for several weeks, but the results are likely to be both informative and novel.

Hybridisation between smooth and palmate newts is of particular interest. Until the 1980s it was believed that these species never hybridised in the wild. Then a male with intermediate morphology was discovered in mid-Wales (Griffiths, Roberts and Sims, 1987) and confirmed as a hybrid by genetic analysis. Since then several more have been found, including one in my own garden pond. Further genetic studies showed that crosses can occur in both directions (male smooth x female palmate and male palmate x female smooth). Several questions arise from these observations. How often does it happen? Is hybridisation promoted by particular environmental conditions? So far only male hybrids have been reported but this is quite likely to be because females of the two species are so similar that it would be hard to spot a hybrid.

The chances of finding a hybrid in the wild are remote, but studies in aquaria could prove interesting. The trick is once again to start off with virgin females that have not yet encountered any males of either species. In aquaria they can then be exposed to males of either the same or the other species, for varying times. How readily will males court the wrong species? How often will females accept them and take up spermatophores? Do hybrid matings differ from normal (control, same-species) matings with respect to the number of eggs that are laid and the proportions that hatch? Are these outcomes affected by the direction of the cross, or other variables such as temperature, or duration of deprivation of same-species partners? There are many possible permutations or extensions of these experiments. What happens if a male of the same species is placed in the same aquarium as a female with males of the 'wrong' species, but separated from her by mesh so that they cannot meet? Will she detect his presence and delay acceptance of the

Hybridisation

Breeding between two different species

'wrong' males? With patience it is possible to rear newt larvae in aquaria, providing them with Daphnia as a food source and thus allowing comparison of growth rates and survival to metamorphosis of hybrids and parental species.

Aquatic habitat selection

A lot has been published about both aquatic and terrestrial habitat features that are important to British amphibians. There is even a 'habitat suitability index' for great crested newts (Oldham *et al.*, 2000; and available for download via www.arguk.org). Once again, though, we still need more information relevant to conservation. Common toads generally do well in ponds with fish, but great crested newts do not. But what about the specifics? Both of these species have declined substantially in recent decades and it would be helpful to know about the effects of particular species of fish. Surprisingly, this topic has scarcely been investigated at all. Are there any species of fish that eat toad tadpoles, distasteful as these are to most (but not all) vertebrate predators? And do all fish species impact adversely on crested newts, or are some tolerated better than others? Unfortunately, garden ponds aren't much help in this case because neither amphibian species is common in them.

Fieldwork to address this question could start by selecting a range of ponds, some of which definitely have fish and some of which don't. There would be a good case for teamwork, which could be achieved by contacting helpers from a local Amphibian and Reptile Group. The bigger the (pond) sample size, the better. Angling clubs will often help with identifying fish ponds and are usually happy to allow research, but of course you will need to check. Local ARGs and/or Wildlife Trusts have general information about ponds, including amphibian species already known to use them. The simplest first step is to look for correlations or associations (positive or negative) between amphibian species and the presence of various species of fish. Angling clubs will usually know what types of fish are in their ponds, but in other cases you may need to find out yourself. To do this you can search with a large pond net and/or with a powerful torch at night as with amphibian surveys, use big versions of the funnel traps (see **4.2**; very effective for sticklebacks) or just go fishing with rod and line (a fishing licence may be needed). You might even persuade the Environment Agency to have a go at electrofishing. There are many guides available for the identification of freshwater fish (such as Maitland and Campbell, 1992). Amphibians can be surveyed as described in chapter 4. Although common toads and great crested newts are of special conservation interest, valuable information about the other amphibians will also accumulate in this type of study. Do smooth and palmate newts occur in a restricted set of microhabitats, such as shallow bays or reed beds, when fish are present?

Life on land
Amphibians are easy to find when they assemble in ponds to breed but they are much less accessible at other times of year. Yet for effective conservation we need to know about life away from water, because that's where all our native species spend most of their time. Where do they go in summer? Where do they hibernate? How many survive from one year to the next? All of this information is vital because habitats the animals use when away from the breeding site need protection just as much as the ponds, if a population is to survive and thrive.

It takes a lot of time and effort to identify terrestrial habitat requirements; that is why there is still a paucity of information for most species. Garden studies are of limited value because most animals range far beyond the confines of one property. In the wider countryside there is a better chance of exploring options more fully but methods are limited. Sets of artificial refugia can be laid out in different habitat types and at various distances from the pond, then inspected at intervals through the summer and autumn. Short sections of fence with associated pitfall traps (see **Fig. 5.6**) can also be deployed. The traps require regular inspection early in the morning to minimise escapes (newts are astonishingly good climbers) and predation by canny predators such as crows, which quickly catch on to easy options for a regular food supply. Torchlight searching at night for foraging amphibians is hard work and only productive on very open habitats such as close-cropped turf. It works effectively with natterjack toads, which like that kind of terrain, but much less well for the other species. Foraging only occurs under favourable weather conditions and these may occur on only a small fraction of the total nights available in summer. Suitable weather for newts to hunt is not necessarily favourable for finding them. It could be interesting to correlate amphibian foraging activity in gardens (they're easy to see on lawns at night) with weather conditions in a rigorous scientific way. This doesn't seem to have been done for any of the widespread species.

Hibernation sites, mostly on land, are also of great interest but are usually discovered by accident rather than design, unless animals are radio-tracked in autumn (an expensive method not available to amateurs). Any you do find, maybe while digging the garden or unloading a compost heap, are well worth reporting as a 'natural history note' (see **9.7**). But what about hibernation under water? Garden ponds can be informative here. Common frogs sometimes overwinter in ponds.The other native species do this much more rarely. Inspecting ponds throughout the winter with a powerful torch, on mild nights, often reveals animals choosing this option because they may become active whenever the temperature rises. Which species can you find in December or January? How many individuals stay in the pond relative to the numbers that breed later in spring? Are pond hibernators males, females or both sexes? How do amphibians in ponds cope with extremely cold weather? Occasionally they don't, and there are regular reports of mass mortalities of frogs floating to the surface after ice melts. Indeed Pond Conservation has run surveys to try to ascertain the frequency of these deaths and to see what, if anything, might be done to minimise the risk in garden ponds. Are big/deep ponds safer than small/ shallow ones? Can we alleviate the problem by removing dead leaves in autumn, so that they won't rot down in winter and deplete oxygen? It is not clear whether the deaths are due to asphyxia or to poisoning by gases such as carbon dioxide or hydrogen sulphide released during decomposition. There is still much to discover about the fate of amphibians in harsh winter weather, and all records are potentially valuable when opportunities arise to make relevant observations. Some experimentation is possible. Will amphibians surface to holes made in ice in order to breathe? If so, does the size, shape or position of the hole matter? How does making a hole affect oxygen concentration in the water beneath ice? Surprisingly little has been done to address this question in garden ponds, and meters for measuring oxygen levels in water are now available at a

Radio-tracked
Followed using a radio receiver after attachment to an animal of a radio transmitter

reasonable price. Attached to data loggers for continuous recording, they could provide fascinating information about what goes on in ponds through the winter months.

Although aspects of amphibian ecology on land are not easy to investigate, information on them would be hugely valuable for conservation. We need to know more about which types of terrestrial habitats are of prime importance, and how near the breeding ponds they need to be. Detection rates are inevitably low away from ponds but fragments of information from numerous observers can help to build the picture. A sound starting point, giving the best chances of detection of the animals on land, is to search habitats around a pond with as large a breeding population as you can find.

Migration

Amphibians are renowned for long and dramatic migrations to their breeding ponds in spring. A great deal has been learnt about the distances involved, the animals' speed of movement and how they find their way. However, an issue of current concern is mortality caused by traffic on roads wherever these cross migration routes. For decades volunteer 'toad patrols' working at night all over Britain have tried to collect (mostly) common toads as they arrive at the roadside and then carry them across. Surprisingly, in light of this very considerable effort by so many people, there have been few attempts to assess the significance of this road mortality on toad populations or the impact on it of help from toad patrollers. Cynics point out that there is nobody to take toads back across after the breeding season when they wander away from the ponds, or to help toadlets when they disperse after metamorphosis. However, road mortality might be critical to the survival of toad populations in at least some situations (Cooke, 2011). If there is a toad patrol active in your area (your local ARG or Wildlife Trust will know) it would be useful to record numbers of toads killed each year, numbers carried across the road and numbers arriving at the breeding site. Are there any correlations over time? If photographic identification can be made to work for common toads the research could go further and assess how many individuals at the breeding site are survivors from a road crossing compared with toads arriving from some other direction (and therefore without ID). In a few places, tunnels to facilitate amphibian migrations have been constructed under main roads using methods and materials developed in the 1980s (Langton, 1989). These tunnels are expensive to make, and it would be of great help in assessing their value to have more data on the numbers using tunnels compared with those trying to cross the road, all in comparison with numbers at the breeding site. Toad numbers have declined in many parts of England over recent years and the reasons for this are not clear. Road mortality could be an important factor and it is vital to find out whether this is so, and thus whether protective measures during migration need extension or improvement.

Longevity

All our native amphibians can live at least into their second decade, and some survive much longer when kept in captivity. In the wild, unsurprisingly, they don't do so well. Studies, mainly using skeletochronology (see **3.2**), have demonstrated this very clearly in a few select populations. However, photographic ID offers the prospect of a wide range of studies based on monitoring individual survivorship over many years. Garden pond populations are ideal for this. Does a newt first photographed in year one return in years two, three and so on? A female smooth newt with a large and distinctive black spot on her back returned to my garden ponds for seven consecutive years. This was a lucky observation before the days of digital photography opened up the possibility of recording more subtle individual variation and thus recording survivorship on a much larger scale.

Introduced species

Considering the success of some non-native amphibians since their arrival in Britain, it is surprising how little has been done to assess their impact on our native fauna. A few species, primarily those from distant continents (American bullfrogs, African clawed frogs), have engendered so much concern that attempts have been initiated to exterminate them, even without direct evidence of any bad effects. Given that both these species are large predators and that our native species have no evolutionary history of living with them, this is surely a wise precaution. The most successful alien invaders, however, stem from mainland Europe where they frequently live together with the British native species. Perhaps that is why there has been less concern about alpine newts and water frogs. But is this relaxed attitude justified? It's true that early worries about marsh frogs ousting common frogs have not been realised, but it would be nice to have some definitive information about how the newcomers have settled in. So if you live near any non-native amphibian populations, there is work to be done.

Alpine newts are widespread but they are still mainly confined to garden ponds, so opportunity for study is limited unless they happen to occupy yours! But if these newts do live in your area there may be an opportunity to find out whether they are associated with habitat features (aquatic or terrestrial) in any way different from those preferred by our native newts. There is some evidence that alpine newts are particularly good at surviving with fish, particularly sticklebacks. More evidence on that issue would be welcome. Where alpine newts share breeding ponds with native newts, as is usually the case, how large are their populations compared with those of native newts? The fact that they all live together in mainland Europe doesn't mean they can necessarily do so equally well here, in a different climate zone. In my garden ponds alpine newts and all three native species have been present since 1978, but great crested newts have gradually declined in recent years. Alpine newts are now the most abundant species. Does this imply slow but damaging competition? Or is the crested newt decline unrelated? More evidence, from more sites, is clearly needed.

Water frogs are now widespread and locally abundant, mostly in countryside ponds and ditches and especially in the south and east of England. So far there is no evidence they have harmed anything and it may be that they are simply occupying an ecological niche previously vacant in Britain. After the end of the last Ice Age, when northern Europe was recolonised from the south by animals and plants that survived the cold there, it's likely that some weren't quick enough to reach Britain before it was cut off by sea-level rise about 8,000 years ago. Maybe water frogs were among these slow-coaches because they need warm summers although we now know that at least some pool frogs did make it in time. But the situation today is that, except in one small corner of Norfolk, all the water frogs encountered in Britain stem from introductions by humans over the past 200 years. These are also among the species most likely to respond to climate change. Warmer summers improve the survival of their tadpoles through to metamorphosis, and after a century or more of barely hanging on in Britain, they have done better within the last few decades and have begun to spread into new habitats (Wycherley and Anstis, 2001).

This means that the simple operation of recording water frogs (preferably identified to species, see **8.1**) at a site and reporting to ARC and/or your local recording scheme is important because it will contribute to an overview about which of them are spreading, how fast, and into what types of ponds. The need to know more about the diet of water frog tadpoles has already been mentioned. But what aspects of habitat are important? In parts of southern England where marsh frogs have become abundant, some ditches have more frogs than other ditches nearby. What features make for an attractive pond or ditch? Water frogs are relatively easy to find and count throughout the summer when they are basking in bright sunshine, always near the water. Maybe some types of vegetation support more invertebrate food or allow the water to warm up more. Fish are often present at water frog sites but which species are tolerated? Are some avoided? And which native amphibians co-occur with water frogs? Does ditch preference change through the year? Perhaps optimal conditions for hibernation (often under water for these frogs), breeding and summer feeding are not the same. Another interesting question relates to segregation of adults and juveniles. Are some ponds or ditches preferred by young frogs and others by adults?

I hope it's clear from this chapter that there are almost endless opportunities for adding to the fund of knowledge concerning amphibians in Britain, despite the work already done by professional researchers over the years. Contributions can vary from very simple to quite complex but even the latter are within the scope of a dedicated amateur enthusiast with time to spare. Relatively little expense is involved and many of the topics suggested would provide information that should improve conservation prospects for these fascinating but threatened animals.

6 Studying reptiles

Unlike amphibians, British reptiles do not normally congregate in specific places to breed. There is an exception to this rule: female grass snakes sometimes assemble at particularly good egg-laying sites such as compost heaps. In all other cases reproduction occurs as widely scattered events in the same habitat that is used throughout the year for hunting and feeding. Furthermore there are no early life stages comparable with spawn and tadpoles in amphibians, that can be investigated separately from adults or juveniles. Both these features of reptile life make study of wild lizards and snakes a more challenging proposition than working with amphibians. On the plus side, all our native reptiles are almost exclusively diurnal. They appreciate much the same sort of weather as humans do, sunny and warm but not too hot, so work conditions are pleasant enough and it is not necessary to keep unsociable hours.

Fieldwork with reptiles requires little in the way of equipment. Binoculars are useful for spotting basking animals several metres away, before they take evasive action and disappear into surrounding cover. With or without these aids, locating lizards and snakes is a skill that benefits from experience. For most people there is a steep learning curve, firstly identifying the places where reptiles are most likely to bask and then picking out their often cryptic colour patterns against the background vegetation. If animals need to be caught they can be captured by hand or, in the case of lizards, with a nylon noose (technique, **9.2**). And as recounted in chapter 4, artificial refugia can be positioned to attract those species that are inclined to use them (mostly snakes and slow-worms). As with amphibians, a good digital camera is a valuable asset for recording colour or scale patterns for future identification of individuals. In a few cases basking animals can be photographed *in situ* without further disturbance. Colour patterns in sand lizards, for example, allow individuals to be recognised in this way.

The options for studies in captivity are more limited for reptiles than for amphibians, whose early life stages can be kept in small aquaria. In general fieldwork rather than captive-based research is recommended, but if it is deemed necessary to keep animals you will need vivaria. These have to be relatively large, even for lizards (**Fig. 6.1**)

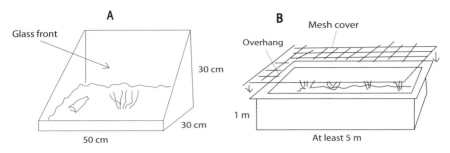

Fig. 6.1 Examples of vivaria for keeping lizards or snakes; A, indoor, small; B, outdoor, large

if there is to be any hope of observing natural behaviour. All our native species (except adders, **9.2**) can be kept quite easily in vivaria, but they need a lot of space to be happy. It is important to position the vivarium so that it is exposed to sun for part of the day, but not so much as to generate overheating. If you have a garden, an ideal option is to build one or more large outdoor vivaria. It is not very costly to create enclosures of several square metres. To prevent escapes these should have vertical walls a metre or so high, dug 20–30 cm into the ground and with internal overlaps at the top (**Fig. 6.1**). They should be planted with suitable vegetation, such as heathers, leaving plenty of bare ground, and the entire edifice should be covered with mesh to prevent predation by birds or cats. Outdoor vivaria are suitable for intensive observation of behaviour throughout the year, although replication is rarely practical because they need to be so big. Large outdoor vivaria have been widely used by conservationists to maintain captive populations of reptiles, particularly sand lizards, which breed freely and produce offspring for release back into the wild.

Feeding reptiles in captivity requires a bit of thought. Limbed lizards thrive on crickets or mealworms, which are readily available from commercial suppliers, but if confined under glass (as in small vivaria) they may need vitamin supplements, dusted onto the insect prey, to compensate for the UV-filtering effect of glass covers. Alternatively, vivaria can be equipped with artificial lighting that mimics natural wavelengths and thus reduces the need for dietary supplements. The British Herpetological Society is a useful source of advice. Although lizards kept in glass-covered vivaria without extra vitamins generally survive, they may fail to produce viable eggs or sperm. Slow-worms require a steady supply of earthworms or, much better, small slugs or snails. Snakes are the most challenging. Grass snakes can often be induced to take dead fish (even fish fingers!), but some may refuse anything but live prey, and frogs are their victims of choice. Smooth snakes are difficult because if individuals of different sizes are kept together they have a propensity to cannibalise. Their preferred food is lizards, which are not easy to obtain in numbers even if you are prepared to do it. Having said that, individuals frequently become tame very quickly and feed readily in captivity, even accepting food offered by hand. Nevertheless, smooth snakes are rarely kept, even for conservation purposes, and of course (as with sand lizards) a licence is required from Natural England (**10.1**). Adders are out of the question for most people, for obvious safety reasons, and to comply with the law they may only be kept when rigorous conditions are provided (see **9.1**). Adders often refuse to feed when confined in captivity unless they have a lot of space, so large outdoor vivaria are a prerequisite. However, they may be induced to accept dead prey such as mice, which can be obtained frozen from commercial suppliers. There is probably much to be learnt about adders by keeping them in outdoor vivaria by anyone with the determination to do so.

6.1 Eggs and juveniles

Sand lizards, wall lizards, grass snakes and Aesculapian snakes lay eggs, while the other reptiles found in Britain all produce live young (see **2.7**). Because sand lizards are of such high conservation interest, much is known about their choice of egg-laying sites and the time taken for their eggs to develop and hatch. However, it would be interesting to know more about factors that influence sand lizard egg survival. This probably depends mostly on weather conditions in summer. Females bury their eggs in sand, usually at night, and cover them up very well, so predation is thought to be minimal. But there could be mileage in relating numbers of hatchling lizards seen in late summer to weather conditions over the critical period from May through to September. Double-clutching is common in sand lizards. Do early clutches, which hatch by mid August in an average year, produce more offspring than late ones that hatch in September? Does the relative success rate vary between years, and does it correlate with climatic factors? It has been postulated that drought and/or unusually high temperatures increase mortality through egg desiccation, and that wet summers cause eggs to succumb to fungal attack. Maybe there's a set of optimal 'goldilocks' conditions between weather extremes that maximise egg survival. It would be useful to know more about weather effects, to help predict how sand lizards will respond to future climate change. If you are lucky enough to live near a sand lizard population, sustained fieldwork and records from the nearest weather station (available via the meteorological office) could contribute valuable information on this subject.

The introduced wall lizard is still rare in Britain and it would be interesting to find out more about how it copes with life at the northerly edge of its range. As far as we know, wall lizards in the UK lay their eggs under vegetation or flat stones, but actually we know very little. So again, if you live near a wall lizard colony careful observation and investigation of possible sites from July onwards could be rewarding. The same is true of Aesculapian snakes but these are so rare that you are unlikely to see one, let alone find their eggs. Grass snakes, on the other hand, are widespread across much of England, but there are still surprisingly few records of natural egg-laying sites as opposed to the well-known artificial ones provided by compost and manure heaps. Any records of natural sites are worth reporting, with full details. Piles of rotting vegetation that generate heat to sustain incubation are the most likely candidates - but who knows? Female grass snakes may travel several kilometres from the ponds and ditches where they hunt to a favoured egg-laying site, so finding them will never be easy and will probably always require a strong element of luck. Even if you find an artificial site, however, it is still worth investigating. Laying refugia around it may give an idea of both the numbers of females attending the site and, later, the numbers of young hatching successfully. Digging into the site in autumn, after hatching is complete, should reveal husks of hatched eggs and the remains of any that died during development. All this information will add to our scant knowledge of the dynamics of grass snake reproduction and help improve conservation measures designed to create new breeding places, which may limit abundance or even the distribution of this species in Britain.

Juvenile reptiles normally appear in late summer or early autumn, whether from eggs or in live-bearing species. Their small size makes them hard to see while basking but for at least two or three days the offspring from the same clutch often stay close together. More observations of this early post-birth period would be useful. Do female live-bearers select particular micro-habitats in which to deposit their young? Are we missing something when carrying out habitat management by ignoring this critical time? There's also a lot still to learn about the influence of weather conditions on the timing of hatch and birth, just as described earlier for sand lizards. As far as we know none of our other reptile species has more than one clutch per year. Is that true? Does bad summer weather delay parturition? If so, what type of weather is important (temperature, rainfall or both?) There are cases on record of young smooth snakes killed by frost in November, perhaps because they were born unusually late. There's also some suggestion that slow-worms may retain their young and give birth the following spring. Does this really happen? More data could provide insights into what might limit population sizes and how future climate change is likely to help or hinder all our native reptiles.

Keeping and rearing hatchling lizards and snakes in captivity is not easy but it could generate interesting information. It would require vivaria as described above, with a sandy substrate including a gradient from damp to dry, and moss or other cover in which the animals can bury themselves for hibernation. Hatchling crickets or sweep netting hedgerows can provide food for small limbed lizards, but there needs to be plenty of it. We don't know much about what very young slow-worms or snakes eat in the wild so there is scope for experimentation here, perhaps with tiny slugs, worms and snails (predation experiments with invertebrates are perfectly legal!). It may be that some individuals do not eat at all before their first hibernation, relying on resources from the embryonic yolk sac to tide them over. Vivaria should be placed in cool (but not freezing) surroundings (maybe in a shed or garage) over the winter. As with amphibians there is an option for using digital photography with these early life stages to find out whether any patterns (scale or spot) are individual-specific when so young and if so, how much they change during growth. This information could be very valuable for future field studies, especially if any patterning is conserved.

6.2 Adult lizards

What are the outstanding questions in lizard ecology and conservation that amateurs might address? Viviparous lizards are a case in point. Some observers have noticed sharp declines in populations of this species in recent years, particularly in southern England, while others have not. Indeed, the opposite may apply in Scotland where this species seems to be doing fine. In the Netherlands viviparous lizards have decreased dramatically over the past decade, more so than any other reptile; this was a considerable surprise (Zuiderwijk and Janssen, 2008). We urgently need more extensive information about the situation in Britain. Eventually this will come through ongoing monitoring efforts as described in chapter 4, but that will take many years to give a clear result. Given the hint that something is awry, immediate and intensive study of specific viviparous lizard populations is highly desirable. Patrolling local habitats to detect basking lizards will reveal the extent of a population and the types of place the lizards inhabit. Is the habitat changing, maybe becoming more overgrown and restricting access to the sun? Exactly what kind of

Fig. 6.2 Viviparous lizards with original, broken and regenerated tails (Chris Gleed-Owen and Fred Holmes)

microhabitats do adult and juvenile lizards use? Broken tails can be a good indicator of predation risk (Cooper, Perez-Mellado and Vitt, 2004). Does the frequency of broken tails vary from place to place? Regenerated tails look quite different from original ones and with practice the distinction is easily made in the field even when there has been extensive re-growth (**Fig. 6.2**).

Sand lizards have a very restricted distribution in Britain but nevertheless have been the subjects of extensive study, and in most cases a licence is required to work on them (certainly if handling is involved). So is there anything left to do? Yes, there certainly is. As we have seen, we need more information on hatching success of eggs and factors influencing this critical mortality rate, but what about adults? One interesting question hinges on the use of marginal habitats. We expect to find sand lizards on mature, dry heathland or coastal sand dunes, and that is where most people look for them. But what about adjacent habitats, especially when these lie between patches of good-quality heathland? Genetic evidence suggests that movement beyond prime habitat must occur at least occasionally and is desirable to prevent population isolation and inbreeding. So what types of suboptimal conditions allow sand lizards to survive, and maybe act as migration corridors? Patient scrutiny of hedgerows, roadside verges, open fields, even parks and gardens could prove profitable. These movements are probably rare and may be mostly carried out by juveniles, so the best chances of discovery will be near large sand lizard populations thriving on optimal habitat. Any such observations will be disproportionately important because there is no understanding of how to maintain such movement corridors. The possible consequences of climate change add further interest. Sand lizards have been increasing in the Netherlands in recent years, possibly because the climate is becoming more favourable for them (Zuiderwijk and Janssen, 2008). In central Europe these lizards occupy a wide range of habitats and are not confined to sandy places. Will we start to see sand lizards spreading into marginal habitats in Britain? Observations around the periphery of established populations are likely to give the first clues of any such shift.

Wall lizard colonies are now quite widespread in southern England. There is concern that they may compete with our native species, especially sand lizards

(Mole, 2010), but we know very little about the ecology of wall lizards in Britain. There is an opportunity to investigate this if you happen to live near one of these introduced populations (see **8.5** for distribution maps; Amphibian and Reptile Conservation will provide exact site details on request). What limits their distribution? How far do they spread from a site of introduction, and what limits this spread? Detailed field studies determining the microhabitats used by all life stages including egg-laying sites (see **6.1**) could provide valuable insights. How important are walls, cliffs and other vertical features? What other reptiles occur in the same habitats? Do these use separate niches (maybe different types of basking areas) or do they overlap? Even casual observations, particularly of behaviour, could be illuminating. Are wall lizards aggressive towards other lizards? Do common or sand lizards living near or among wall lizards have more instances of broken tails than those living away from the intruders? Any and every bit of information would help us to evaluate how worried we should be about the spread of this otherwise attractive and lively addition to the British fauna.

The enigmatic slow-worm has been the subject of precious few studies, although it is one of Britain's commonest reptiles. This is largely due to its secretive behaviour. Indeed, over the past forty years there have been approximately three times more papers in the scientific literature on viviparous lizards, and twice as many on sand lizards, as on slow-worms. Yet despite the difficulties of finding them, the predilection of slow-worms to live and thrive in gardens offers especially rich opportunities for amateur study. Once again there are potential conservation benefits. Because this species is so adaptable to urban environments it features regularly in mitigation measures required as a condition for acceptance of development proposals (Platenberg and Griffiths, 1999). Usually slow-worms are caught and moved to a new site, but how useful is this without sound knowledge about their requirements?

Much very basic information could be gained in gardens just by setting out refugia (see **4.3**) and monitoring the animals that use them. Although slow-worms have few obvious marks for recognition, digital photography of scale patterns around the head has been used with some success to monitor individuals. Other markings can also be distinctive; spots which can change between blue and brown occur on some individuals, particularly males, which may also bear distinctive scars from fights with rivals during the breeding season. There is scope for more studies along these lines, perhaps following animals as they grow to determine how well individual patterns are conserved during development to adulthood. Home ranges are likely to extend beyond individual gardens, but this does not preclude valuable research. Which microhabitats within gardens are preferred (flower beds, shrubberies, uncultivated patches, compost heaps)? What environmental conditions trigger daytime hunting? In my garden I occasionally see slow-worms actively moving around, usually on summer evenings, presumably looking for food. But this behaviour is rare and probably dangerous, because slow-worms are preyed upon by birds and cats in urban gardens. Is this risk-taking promoted by particular weather, maybe bringing out favoured prey like small slugs?

Regular records of slow-worms under refugia should generate useful information on other basic aspects of their ecology and behaviour. How frequently can they be disturbed without causing them to move somewhere else? How much movement occurs within a garden at different times of year? How do numbers of males, females and juveniles vary month by month between emergence from and re-entry to hibernation, and how does the hibernation period vary according to weather conditions? Is there any evidence of females giving birth in spring? Long-term studies would also be interesting. Do males and females survive equally well? Slow-worms in captivity can be astonishingly long-lived (with over fifty years recorded on one occasion) - but how do they fare in the wild? It is even possible to find out what slow-worms have been eating. Keeping individuals at 8°C (in a fridge) for up to five days causes about half of them to regurgitate their food. This can then be identified, and the lizards released unharmed (Pedersen, Jensen and Toft, 2009). It would be intriguing to find out how diet varies across the year in different age and sex groups, both qualitatively and quantitatively.

6.3 Adult snakes

Snakes offer rather few opportunities for amateur study, for several reasons. Adders are problematic because of the bite risk, and smooth snakes are so rare and secretive that they are out of the question for most people. All snakes are higher in the food chain than our other herpetofauna with the consequence that population density is relatively low and individuals often have large home ranges. There are, of course, no snakes in Ireland. But they are fascinating animals, and given sufficient determination and enthusiasm there are some prospects for interesting work.

Grass snakes are the best bet for amateur study and are widespread in much of England and Wales, though essentially absent from Scotland. In gardens you may be lucky enough to have grass snakes visiting a pond to catch prey, or a compost heap in which to lay eggs. What are they after in garden ponds? Do they go primarily for fish or are they focusing on amphibians? In the wild, grass snakes seem to prefer amphibians, perhaps because they are the more easily caught. If this is also true in gardens, the knowledge would give useful reassurance to pond owners, thus reducing any persecution the snakes might otherwise suffer. Simple observation, taking care not to disturb or frighten the snakes, might provide some answers. Otherwise, a grass snake that has recently eaten something (there will be a conspicuous bulge in its body) will often regurgitate its last meal if it is placed in a cloth bag for an hour or so. This regurgitation can even be encouraged, with no harm to the snake, by gently squeezing below the bulge. Don't squeeze hard though; if nothing happens quickly, give up. The results can be a little gruesome. Once again digital photography to identify individuals (head scales, collar markings and so on) will make it possible to estimate the numbers of individuals visiting a garden pond or compost heap, how often they come, and for how many years they re-appear. This would all amount to valuable and novel contributions to grass snake ecology.

Study needn't be confined to garden visitors if you know of an area, perhaps with a pond or marsh, that is regularly frequented by grass snakes. Just as slow-worms occasionally show unusual behaviour by hunting in daylight (see above), large numbers of grass snakes sometimes bask overtly in exposed places in contrast to their more usual thermoregulation under partial cover or refugia. I've walked round a pond with grass snakes lying out along the bank every few metres but this seems to be a rare event. On other warm days I've seen few or none at all at the same pond and wondered why. What triggers this behaviour, with its (presumably) increased risk of predation? Regular visits to a suitable pond with a large grass snake population should reveal how common this is, whether it's confined to particular times of year and whether it correlates with specific weather conditions. A better understanding of the triggers of such conspicuous behaviour would be helpful for informing monitoring programmes, and might even provide a way of estimating population size.

Adders merit attention despite health and safety considerations which should always be taken carefully into account. Handling these snakes should be left to experienced herpetologists. However, useful fieldwork can be achieved without significant risk, and in Scotland this is the only species of snake available for study. Unlike grass snakes, which range freely across many agricultural landscapes, adders are restricted in the types of habitat they occupy. This means they are more vulnerable to habitat fragmentation, and many populations are increasingly isolated in preferred sites such as well-drained scrubby slopes, heaths and dunes. In Sweden such isolation has led to inbreeding depression (Madsen, Stille and Shine, 1996) and this may be happening in Britain too, particularly in lowland England where farming is most intensive and where adder populations have declined dramatically in recent decades. As with sand lizards, we need to know what constitutes suitable corridor habitat to maintain connectivity between these relict populations. So if you find good adder habitat with numerous snakes, explore the periphery by putting down refugia. Check out roadside verges, hedgerow edges, indeed pretty much anywhere exposed to the sun where adders might go. Note the microhabitat details where you find snakes and, just as important, where you don't find them: record plant species and height of vegetation, soil moisture and type, distance from the main population and so on. These data could be enormously helpful in planning future conservation strategies aimed at keeping adder populations linked together by occasional migration.

Fig. 6.3 A fenced site for reptile removal studies (Tim Harry)

6.4 General questions about reptiles

Aside from the kinds of investigation outlined above there are some more general questions of conservation importance which relate to all of our native reptiles. A couple of examples potentially amenable to amateur study are described below. Both are a bit more complex and expensive than the earlier suggestions but are certainly possible for the keen enthusiast.

Mitigation measures imposed on developers by planning authorities frequently involve moving reptiles, especially viviparous lizards and slow-worms, away from a proposed development site. Usually this entails multiple visits by a contracted consultant to fence the area and collect the animals before work starts. However, there is still remarkably little information on the efficiency of this protocol. How many visits does it take to catch 50%, 75%, 90% or more of the reptiles present? How does this vary according to season? Guidelines at present are little more than guesswork. A widespread view is that they should be more evidence-based and probably stricter and more rigorous than those usually demanded of the developer or consultant. Field experiments could provide this evidence. They would require compartmentalising part of a suitable habitat containing a good-sized reptile population (perhaps on a downland slope) using reptile-proof fencing. The area need not be large (**Fig. 6.3**). The temporary fencing can be made from cheap plastic dug into the ground, with wooden supports and ideally with an overhang on both sides to prevent escapes or re-entries of moved animals. Then refugia are set out within the site, and visited at regular intervals. Any animals present are caught and released outside the fence. Is a point ever reached when no more animals are found? What is the rate of decline of the contained population? Some studies attempted during mitigation work have suggested that, particularly for slow-worms (the species for which this experiment is best suited) it takes a big effort before diminishing returns become apparent. Can this situation be improved by increasing the density of refugia?

A second and particularly big issue in reptile conservation concerns livestock grazing on habitats with good reptile populations. Cattle and sheep inevitably impact on vegetation structure, perhaps to the detriment of wildlife that is dependent on the integrity of that structure. Grazing has become a hot topic particularly on heathlands which sustain our rarest reptiles (sand lizards, smooth snakes) as well, very often, as sizeable populations of the more widespread species. Mature stands of heather provide optimal habitat, creating cover from predators, good food supplies and easy access to sunshine for thermoregulation. Unfortunately heather is also very sensitive to damage by trampling, which if excessive can obliterate mature stands in a short time. Heathland ecologists nevertheless increasingly advocate grazing by domestic animals as the best approach to long-term, sustainable management of this endangered habitat. They point out that heathlands were initially created by humans several thousand years ago as a consequence of forest clearance and were subsequently maintained mostly by grazing until about a century ago when this became uneconomic. Without constant attention heathlands succumb to invasion by bracken, scrub and trees (especially pine seedlings) and eventually become completely overgrown. They are 'pre-climax' ecosystems and the best way to maintain them, surely, is to revert to the methods our ancestors used successfully for millennia. Reptiles everywhere need 'open' habitats, not just on heathlands, and therefore broadly similar arguments have a universal application.

Unfortunately it is not as simple as that. Heavy grazing is undoubtedly damaging to the whole habitat and it is crucial to ensure that the density of livestock is low enough to prevent wholesale destruction. Because many of the remaining heaths persist only as small fragments, the practicalities of fencing and maintaining the necessarily small numbers of grazers can present potential problems and be less cost effective than keeping heathland open by manual scrub clearance. Because livestock is usually borrowed from a local farmer, conservationists advocating grazing regimes are rarely in a position to choose the breed of animal, their numbers or the time of year when they are on site. All of these variables make a big difference to what happens on the ground.

The debate therefore hinges on whether grazing regimes can be made compatible with effective heathland management (controlling invasive vegetation) while at the same time leaving heather intact enough to support reptile populations. Or on other habitats such as hill pastures, can grazing be adjusted to leave sufficient rank vegetation with patches of cover to sustain the widespread reptile species? There have been some disasters when overgrazing has resulted in savage destruction of habitats and species, and they have polarised debate on the subject for many years. But the possible rewards of improved understanding are high. Appropriate grazing could indeed turn out to be an ideal management tool. Funding for heathland management, provided to non-government organisations by statutory agencies such as Natural England, is now largely conditional on establishing some sort of grazing regime, implying that in many quarters the debate is over. This makes it all the more remarkable that there has been so little research into the impact of grazing on reptile populations. We know that grass snakes tend to avoid grazed fields (Reading and Jofre, 2009) and that smooth snakes declined on one area of grazed heath whereas other reptiles, including sand lizards, did not (Reading, 2010), but that's about it.

No doubt the best way to investigate the impact of grazing on reptiles is by carefully designed experiments comparing population dynamics in areas fenced off from livestock with dynamics in adjacent unfenced areas, but this is expensive and takes many years to assess. However, other approaches are possible which, although less rigorous, could provide useful insights relatively quickly. All that's needed is to find pairs of sites with apparently similar and potentially suitable reptile habitat, one grazed and the other not. They should be as physically close as possible, certainly on the same terrain (downland hillside, moor or whatever) and without any other features apart from grazing that would make one obviously better than the other. For example, slopes should face broadly in the same compass direction. Then set out multiple refugia on each site and inspect them as well as walking an optimal transect line at regular intervals to find basking reptiles (see **4.3**). Over a spring and summer season it should become clear whether the population sizes of any reptiles that occur in the area differ between grazed and ungrazed regimes. Reproductive success is also worth noting by comparing the numbers of juveniles that turn up late in the summer. At the same time as monitoring the reptile populations it is also important to note details of the grazing regime (number of animals per hectare, breed of animals, whether or not grazing is continuous) and to describe the vegetation, at least in terms of average sward height and the extent of other cover such as occasional shrubs or gorse bushes. If even a few people carried out this kind of investigation the results could be hugely valuable for future conservation planning, and would help to fill a vacuum still not adequately addressed by professional researchers.

7 How schools can help

Some of the most important questions in amphibian and reptile natural history are best tackled by a group of people, such as a school class or a natural history society, because the animals are distributed over a wide area and are seldom found in really large numbers in any one habitat. For this reason there are opportunities for collaborative work to make a particularly valuable contribution to the body of information that underpins conservation management of amphibians and reptiles.

Three types of investigation are outlined below. In the first and most universally applicable, the opportunity to sample what's going on in the school's catchment area is highlighted. The second approach is more limited and based on what happens to be available within the school grounds, or maybe in a nearby park or nature reserve that a supervised class might visit. The third is lab-based and provides opportunities for close encounters with these fascinating creatures.

7.1 News from home

A potentially valuable set of related topics exploits the fact that many people live in homes with gardens, and some of these are likely to have ponds. First of all: how many homes have ponds, what percentage of the class does this amount to and are they randomly distributed around the catchment or concentrated in particular areas? Next of course are the obvious questions about amphibians: do you know what visits your pond? Do you ever see frog or toad spawn, or newts? If you don't know, try and find out in the spring breeding season. And when all the records are in, look to see whether the distribution of each species is random or associated with some parts of town but not others. What percentage of ponds is each species found in? Then, how about some more information about those ponds? How big are they, how deep, how old if that happens to be known? Do they have fish and if so, what kind? Are they shaded by trees or exposed to the sun? Do they have lots of pondweed, little, or none at all? The answers to such questions can give useful information on habitats occupied by the different species.

With reptiles there are fewer opportunities because they are less common than amphibians in garden habitats. Slow-worms, however, are a widespread exception. Again, find out which gardens have them. Is the distribution random and if not, what features correlate with their presence? For example do they like particular soils, aspects (maybe south-facing slopes), types of vegetation (degree of tree cover, extent of rough ground left uncultivated)? In some places grass snakes visit gardens regularly and offer more opportunities for class-wide studies. People are often fascinated or frightened by snakes, sometimes in equal measure! Here is a chance to improve the image of these exciting animals. As with amphibians it is possible to map the distribution of garden ponds visited by grass snakes. Do they correlate with big or small gardens, availability of compost heaps, and/or proximity to open countryside? Even better, pupils may provide observations on what the snakes catch when they visit their ponds.

In a short space of time a bunch of enthusiastic participants can provide a large and remarkably interesting data set. Depending on their age they can be encouraged to analyse the results with varying levels of sophistication including statistical methods where appropriate. Maybe the local distribution of a species maps onto a particular feature such as the underlying type of soil, which can be determined by reference to national databases. Perhaps some species of amphibians, or grass snakes, occupy new ponds quickly while others only turn up in older ones. Are big ponds better than small ones? Do fish make a difference and if so, in what way?

7.2 Field study

Possibly a local pond, perhaps even in the school grounds, could become the focus of attention. If so, set out to discover which amphibians breed in it using some of the methods described in chapter 4. Visual inspection is adequate for frog or toad spawn and using a pond net should be fine for newts and tadpoles, with due care (of course) for safety requirements. At the same time make an inventory of other interesting wildlife (fish, aquatic insects and so on). This approach is well suited to continuation over several years, maybe with different classes or with the same cohort as students progress through the school. When is the first frog spawn seen? And the first newts? Do the dates change over the years (perhaps climate change at work) and do they relate to weather conditions each spring? Netting through spring and early summer will give an idea of whether tadpoles are surviving, and look out for metamorphs in June. Does breeding success vary between years? Does it correlate with other things going on in the pond such as the number of dragonfly nymphs you catch in the net? The effects of management could also be investigated if the pond belongs to the school. Does removing pondweed in autumn or cutting down overhanging bushes affect what happens in the following year?

Once again it's easier to carry out school-based fieldwork with amphibians than it is with reptiles. Maybe, though, there is a sunny bank with low vegetation not regularly cut or mown in a corner of the school grounds or somewhere nearby. There's not much chance of a group of children creeping up on viviparous lizards before the animals scuttle away, but slow-worms are once again an option. If the place is relatively undisturbed, small refugia (see **4.2 and 4.3**) could be left in place for a while and then inspected to see what turns up. If slow-worms appear it would be interesting to monitor the numbers of each sex, and to see when young appear, and to record interesting colour varieties through the spring and summer, indeed well into the autumn. Some may have blue spots and very occasionally jet black individuals are seen. How many have broken or regenerated tails? Does this differ between the sexes? Is most new damage seen at specific times of year? What else lurks under the refugia? Because we still know so little about these secretive reptiles, all observations on what they get up to are valuable.

7.3 Lab study

For a long time, frog spawn has been kept and tadpoles have been reared through to metamorphosis in schools as a demonstration of animal growth and development. It remains a very worthwhile exercise provided it is done carefully; tadpoles should not be overcrowded (keep just a few after the spawn hatches), or overfed (the occasional rabbit pellet or boiled lettuce leaf will suffice, so as not to pollute the water), water must be changed if it starts to look murky and any extra tadpoles as well as froglets should be released promptly at the pond they came from. Transport, particularly of froglets, needs special attention. Keep them in damp weed, not in water, because they drown remarkably easily after metamorphosis.

Less often done in schools but also worthwhile purely for observation is the keeping of a few smooth or palmate newts for a week or two in an aquarium during the height of the breeding season (usually some time in April). The courtship rituals of newts are easy to observe and fascinating to watch, with males surging in front of females and vibrating their tails. If some water weed is provided it is also possible to watch the females fastidiously wrapping eggs in the weed leaves. Make sure the tank is covered (newts are astonishing escape artists), and provide food such as Daphnia or Tubifex worms, which are readily available from aquarist shops (see 10.1). Release the adult newts within a couple of weeks; it isn't really fair to keep them away from their main breeding site for longer. If eggs are laid it is possible to rear the larvae (again, only keep a very few) and feed them on small Daphnia or other tiny pond creatures. However newts take much longer to develop fully than frog or toad tadpoles, and may need to be released, still as tadpoles, before the start of the summer holidays.

Lab work with frog or toad tadpoles can help to establish the best conditions for rearing amphibians by generating comparisons. Try rearing tadpoles at different densities (say five per tank and twenty per tank), either with the same total amount of food per tank or per tadpole. Or try the same tadpole densities at different temperatures (warm in the classroom, cooler in an outhouse). Measure how long it takes, in each case, for the tadpoles to start metamorphosis. Many other permutations are possible. How does the amount of food influence growth rate? And what about the type of food? Compare fish flakes, rabbit pellets and tiny pieces of meat (taking care, in this case, that the water doesn't turn foul from decomposing, uneaten remains). And what about water changes? Are growth rates affected by the simple expedient of replacing the aquarium water at different intervals?

There are two different processes going on in a tadpole. One is straightforward growth, but the other is development, meaning the differentiation of body parts such as legs. These two aspects of tadpole life are not necessarily closely correlated. You might notice that it's not just time to metamorphosis that varies according to growth conditions. Tadpoles may grow bigger in some treatments than in others. It's possible to measure the length of tadpoles quickly and safely by catching them in a tea strainer and holding them out of the water, very briefly, against a ruler – then returning them immediately to the tank. At the same time, you can record the developmental stage (see **Fig. 5.3**). Do the biggest tadpoles also develop faster than the smallest? And does the relationship stay the same if the variable is temperature or food supply?

Yet again amphibians are easier to work with than reptiles in lab studies, and in a school environment amphibian spawn and tadpoles are usually the only option. But it is a good option, with lots of scope for imaginative studies. Remember, though, that adding predators in this situation is illegal – it is not appropriate to investigate the effects of dragonfly nymphs or water beetles on tadpole survival in any school-based scientific study.

7.4 Overview

Amphibians must be among the most amenable of subjects for school-based biology. All the types of study described above can make a real contribution to knowledge, because they bring together a lot of information at a local scale. Schools are well placed to undertake such research, and there's no reason why the results shouldn't be published (as described in **9.7**) to produce an outcome that the school can be proud of. At the same time pupils can have the chance to see wild animals and experience the excitement of completely new discoveries based on the simplest of scientific methods.

8 Identifying species found in Britain

There are only seven amphibian and six terrestrial reptile species native to Britain and in most cases their identification is straightforward. However, a further seven amphibian and three reptile species, introduced over the past two hundred years, have succeeded in establishing one or more self-sustaining populations. It is therefore important to recognise these animals as well. On top of that are many others that have been found occasionally or which have established breeding populations for short periods. These species, which are not considered further in this book, include European tree frogs *Hyla arborea*, yellow-bellied toads *Bombina variegata*, fire salamanders *Salamandra salamandra*, marbled newts *Triturus marmoratus*, American bullfrogs *Lithobates catesbeianus*, Italian wall lizards *Podarcis sicula*, dice snakes *Natrix tessellata*, garter snakes *Thamnophis sirtalis*, corn snakes *Pantherophis guttatus*, European pond tortoises *Emys orbicularis*, red-eared terrapins *Trachemys scripta*, painted terrapins *Chrysemys picta*, Hermann's tortoises *Testudo hermanii* and spur-thighed tortoises *Testudo graeca*. There could be more. An excellent well-illustrated guide to these and to our native species is provided by Inns (2009). The following keys together with associated photographs and distribution maps are intended to facilitate identification of all our native species and the non-native ones most likely to be found in Britain today.

Fig. 8.1 Eggs of smooth newt *Lissotriton vulgaris* (Howard Inns)

Fig. 8.2 Eggs of great crested newt *Triturus cristatus* (Chris Gleed-Owen)

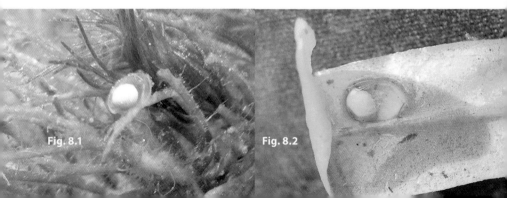

Fig. 8.1   Fig. 8.2

Fig. 8.3

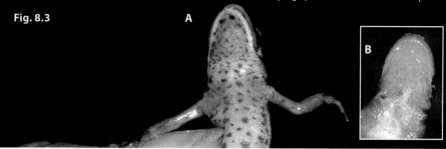

Fig. 8.3 Throats of smooth newt *Lissotriton vulgaris* (A) and palmate newt *Lissotriton helveticus* (B) (Erik Paterson and Steve Ogden)

Fig. 8.4 Larva of smooth newt *Lissotriton vulgaris* (Howard Inns)

Fig. 8.5 Early stage larva of great crested newt *Triturus cristatus* (Charles Snell)

Fig. 8.6 Mid-stage larva of great crested newt *Triturus cristatus* (Charles Snell)

Fig. 8.7 Late-stage larva of great crested newt *Triturus cristatus* (Charles Snell)

Fig. 8.8 Larva of alpine newt *Ichthyosaura alpestris*

Fig. 8.9 Late-stage larva of alpine newt *Ichthyosaura alpestris* (Charles Snell)

Fig. 8.10

Fig. 8.11

Fig. 8.12

Fig. 8.10 Smooth newt *Lissotriton vulgaris*, female (Howard Inns)

Fig. 8.11 Smooth newt *Lissotriton vulgaris*, male (Fred Holmes)

Fig. 8.12 Palmate newt *Lissotriton helveticus*, female (Howard Inns)

Fig. 8.13 Palmate newt *Lissotriton helveticus*, male (Fred Holmes)

Fig. 8.13

Fig. 8.14

Fig. 8.15

Fig. 8.14 Great crested newt *Triturus cristatus*, female (Fred Holmes)
Fig. 8.15 Great crested newt *Triturus cristatus*, male (Howard Inns)
Fig. 8.16 Italian crested newt *Triturus carniflex*, male (Charles Snell)
Fig. 8.17 Italian crested newt *Triturus carniflex*, female (Charles Snell)
Fig. 8.18 Alpine newt *Ichthyosaura alpestris* (Fred Holmes)

Fig. 8.16

Fig. 8.18

Fig. 8.17

Fig. 8.19

Fig. 8.19 Common fog *Rana temporaria* (Fred Holmes)

Fig. 8.20 Native pool frog *Pelophylax lessonae* (Fred Holmes)

Fig. 8.20

Fig. 8.21

Fig. 8.21 Common toad *Bufo bufo* (John Wilkinson)
Fig. 8.22 Natterjack toad *Bufo calamita* (Fred Holmes)

Fig. 8.22

Fig. 8.23

Fig. 8.23 Pool frog *Pelophylax lessonae* (introduced) (Howard Inns)

Fig. 8.24 Edible frog *Pelophylax esculentus* (Fred Holmes)

Fig. 8.25 Marsh frog *Pelophylax ridibundus* (Paul Wells)

Fig. 8.26 Midwife toad *Alytes obstetricans* (Howard Inns)

Fig. 8.27 African clawed frog *Xenopus laevis* (Ronn Altig)

Fig. 8.24

Fig. 8.25

Fig. 8.26

Fig. 8.27

Fig. 8.28

Fig. 8.28 Spawn of common frog *Rana temporaria* (Neal Armour-Chelu)

Fig. 8.29 Spawn of common toad *Bufo bufo* (Howard Inns)

Fig. 8.30 Spawn of water frog *Pelophylax* sp. (Howard Inns)

Fig. 8.31 Spawn of natterjack toad *Bufo calamita* (Angela Reynolds)

Fig. 8.29

Fig. 8.30

Fig. 8.31

Fig. 8.32 Larva of common frog *Rana temporaria* (Howard Inns)

Fig. 8.33 Larva of water frog *Pelophylax* sp. (Howard Inns)

Fig. 8.34 Larva of common toad *Bufo bufo* (Howard Inns)

Fig. 8.35 Larva of natterjack toad *Bufo calamita* (Howard Inns)

Fig. 8.36 Larva of midwife toad *Alytes obstetricans* (Howard Inns)

Fig. 8.37 Larva of African clawed frog *Xenopus laevls* (Ronn Altig)

Fig. 8.38

Fig. 8.39

Fig. 8.40

Native species

Fig. 8.38 Slow-worm *Anguis fragilis* (Chris Gleed-Owen)

Fig. 8.39 Viviparous lizard *Zootoca vivipara* (John Wilkinson)

Fig. 8.40 Sand lizard *Lacerta agilis* (Fred Holmes)

Introduced species

Fig. 8.41 Wall lizard *Podarcis muralis* (John Wilkinson)

Fig. 8.42 Western green lizard *Lacerta bilineata* (John Wilkinson)

Fig. 8.41

Fig. 8.42

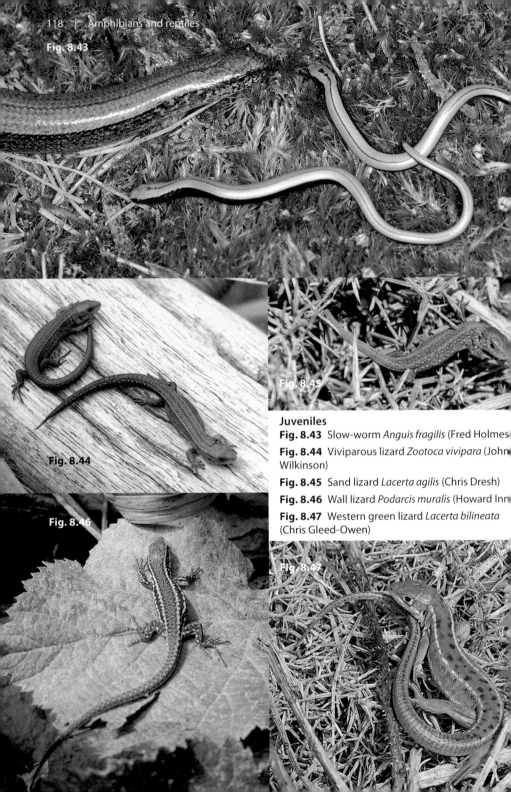

Fig. 8.43

Fig. 8.44

Fig. 8.45

Fig. 8.46

Fig. 8.47

Juveniles

Fig. 8.43 Slow-worm *Anguis fragilis* (Fred Holmes)

Fig. 8.44 Viviparous lizard *Zootoca vivipara* (John Wilkinson)

Fig. 8.45 Sand lizard *Lacerta agilis* (Chris Dresh)

Fig. 8.46 Wall lizard *Podarcis muralis* (Howard Inns)

Fig. 8.47 Western green lizard *Lacerta bilineata* (Chris Gleed-Owen)

Fig. 8.48 Grass snake *Natrix natrix* (Fred Holmes)
Fig. 8.49 Adder *Viper berus* (Roger McPhail)
Fig. 8.50 Smooth snake *Coronella austriaca* (Howard Inns)
Fig. 8.51 Aesculapian snake *Zamenis longissimus* (Wolfgang Wüster)
Fig. 8.52 Grass snake *Natrix natrix* with eggs (Neal Armour-Chelu)

8.1 Amphibians

The British amphibians all fall within two easily distinguished orders: the tailed species (newts) and the tail-less species (frogs and toads). Keys for each order are given separately below.

Order

Taxonomic grouping above the level of genus, which in turn is above the level of species

Newts are most frequently encountered in ponds but at other times of year when on land and hiding under refugia they may at first sight be mistaken for lizards. Newts invariably have a smooth, often velvety and/or damp skin and move slowly (or not at all!) when disturbed. Lizards always have scaly, dry skin and move fast to evade capture. The following key covers five species of newts, three natives and two introductions. Some species are not distinguishable at the egg or larval stages.

All amphibians in Britain either retain tails after metamorphosis (urodeles: the newts) or lack tails after the tadpole stage (anurans: the frogs and toads). Three life stages can be distinguished in both groups.

1.   A larva is entirely aquatic, and always has a finned tail.

2.   A metamorph is a young amphibian in the weeks immediately after metamorphosis, up to its first hibernation. It generally resembles a miniature version of the adult in both urodeles and anurans, though colouration sometimes differs.

3.   An adult or immature is an animal in or after the second year of life. Immatures are similar to adults in most respects, but sex cannot usually be determined before maturity (in most cases immatures have female morphology).

In the following keys, the diagnostic contrasting features come first, followed by confirmatory features (not necessarily different in the two leads) in brackets.

Key I Adult and immature newts

Immatures and adults are identical except for size and are treated together in this key.

Sexing newts

Male smooth newts and male crested newts have impressive crests in the breeding season. Crests regress in summer, although traces usually remain visible. Apart from the crests, the size of the cloaca is a good guide to sex. Male cloacas in all species are substantially larger than those of females all year round. It is not generally possible to sex immature newts (those less than two or three years old); they all look like females.

1   Small (less than 10 cm total length); (upper body mostly brown, sometimes with dark spots; there may be a distinctive wavy crest along the back and tail in the breeding season)   **2**

-   Larger (more than 10 cm long); (upper body dark or even black; there may be an obvious crest along the back and tail in the breeding season)   **3**

2   Throat pink and unspotted (**Fig. 8.3 B**); (belly pale yellow with few or no spots; breeding males may have black, webbed hind feet and a short filament at the end of the tail, but no high crest; females usually have an inconspicuous pale bar above the hind limbs)
Palmate newt, *Lissotriton helveticus*
(Fig. 8.12 & Fig. 8.13, Map 2)

-   Throat white, usually spotted with black (**Fig. 8.3 A**); (belly yellow orange, with black spots; breeding males may have a continuous, wavy crest on the back and tail, and a blue stripe along the tail; females have no pale bar above the hind limbs)
Smooth newt, *Lissotriton vulgaris* (**Fig. 8.10 & Fig. 8.11, Map 1**)

It can be difficult to distinguish smooth and palmate newt adults, especially females. Newts with weakly spotted or unspotted but white rather than pink throats turn up occasionally and are usually smooth newts. Palmate newts are rare in East Anglia and smooth newts are rare in Cornwall and in the Scottish mountains. Be particularly careful to check identification of newts from these regions, for example, suspected palmate newts from Norfolk.

3 Large (more than 12 cm long); belly striped or spotted with black on a yellow or orange background **4**

- Medium sized (10–11 cm long); belly unspotted, bright orange; (back dark blue-black, sometimes with green marbling; males have yellow spots on a low crest along the back; male flanks spotted and with a distinctive blue line)

Alpine newt, *Ichthyosaura alpestris* (Fig. 8.18, Map 4)

Alpine newts have, thus far, mainly turned up in garden and park ponds, but are widely distributed in Britain including Scotland.

4 Dark spots on belly well defined; (breeding males have a jagged crest on the back and tail, with a gap in the crest between back and tail, and a silver stripe along the tail; white spots along flanks; usually no yellow vertebral stripe in females)

Great crested newt, *Triturus cristatus*
(Fig. 8.14 & Fig. 8.15, Map 3)

Great crested newts are rare in Cornwall and in the Scottish mountains.

- Dark spots on belly have fuzzy edges; (like great crested newt, but with few white spots along flanks, and a yellow vertebral stripe is usually present on females)

Italian crested newt, *Triturus carnifex* (Fig. 8.16, Map 5)

Italian crested newts are only known from Surrey and near Birmingham. In both places they co-occur with native great crested newts and hybridise with them, generating individuals of intermediate morphology.

Key II Newt eggs, larvae and metamorphs

Metamorphs are small newts found within weeks of leaving the pond, normally in late summer or autumn. Except for smooth and palmate newts, which may have vertebral stripes, metamorphs are otherwise like miniature adults.

1 Egg or embryo, still within jelly capsule **2**
- A free-swimming larva or metamorph **3**

Fig. 8.53 Back stripes on newly metamorphosed smooth (A) and palmate (B) newts (from Roberts and Griffiths, 1992 [reproduced with permission from Brill]). Sometimes stripes do not appear until a few weeks after metamorphosis

A

B

2   Egg or embryo brown or grey-buff; total diameter (including jelly) around 3 mm
   Palmate, smooth or Alpine newt (Fig. 8.1)

-   Egg or embryo white or pale lime yellow or green; total diameter around 5 mm
   Great crested or Italian crested newt (Fig. 8.2)

3   Aquatic larva; with obvious feathery gills sprouting from neck region   **4**

-   Terrestrial metamorph; no gills   **6**

4   Tail clearly spotted; (grey brown, with a high fin tapering to a point; up to 70 mm long)
   Great crested or Italian crested newt (Fig. 8.6)

-   Tail not clearly spotted, but may be speckled; (moderate or low tailfin; not more than 40 mm long)   **5**

5   Uniform light brown; (up to 30 mm long when fully grown)
   Palmate or smooth newt

-   Dark blue or black; (up to 40 mm long; heavily speckled, with a broad, blunt-ended tail fin
   Alpine newt, *Ichthyosaura alpestris* (Fig. 8.9)

6   (Terrestrial metamorphs) 60–70 mm long; (black above; belly yellow with black spots; skin warty)
   Great crested or Italian crested newt

-   Less than 60 mm long   **7**

7   Bluish above; up to 40 mm long; (may be spotted; belly uniform yellow orange)
   Alpine newt, *Ichthyosaura alpestris*

-   Brown (with a pale underside); up to 30 mm long; (may have a yellow vertebral stripe)   **8**

8   Vertebral stripe, if present, starts in middle of head and fades along back
   Smooth newt, *Lissotriton vulgaris* (Fig. 8.53 A)

-   Vertebral stripe, if present, starts on neck and runs down uniformly to tail
   Palmate newt, *Lissotriton helveticus* (Fig. 8.53 B)

8.2 Frogs and toads

The following keys cover nine species of frogs and toads, four native and five recently introduced to Britain. One, the pool frog, occurs in both categories because different native and introduced forms exist in the wild although morphological differences between them are minor, thus making the distinction difficult in practice. Indeed, it is often difficult to differentiate all the water frogs (marsh, edible and pool), and to do this it is usually necessary to catch and handle the animals. Even then the distinction may not be convincingly clear. As with newts in some cases it is also impossible to identify eggs and larvae at species level. Metamorphs are not considered separately because by that stage of development they have adult characteristics and the adult keys suffice. Natterjack metamorphs, for example, bear the yellow vertebral stripe.

Sexing frogs and toads

Males of all species have distinctive calls in the breeding seasons and excepting midwife toads and clawed frogs develop pads on their front feet (often darkly coloured) for clasping females. Males also generally have stouter forearms than females, for the same reason. Throats of male common frogs and natterjacks are bluish/purple, again most marked in the breeding season; those of females have a white background, often with dark spots. But sex determination can be difficult, especially outside the breeding season, and with midwife toads is virtually impossible unless there are eggs around the hind feet (these are only carried by males).

Key III Frogs and toads (adults and metamorphs)

1   Skin moist and mostly smooth, without large warts;
    (upper body colour variable but predominantly brown,
    yellow or green, sometimes with dark blotches)   **2**

-   Skin relatively dry, and warty; (upper body colour
    usually grey or brown, occasionally with greenish
    tinge)   **6**

2   Prominent brown patch behind eye; often far from
    water outside breeding season; (upper body colour
    varying from yellow through to brown, rarely
    green; male throat blue in breeding season; females
    sometimes marked with red spots on flanks; nose
    relatively blunt)
                Common frog, *Rana temporaria* (Fig. 8.19, Map 6)

-   Little or no eye patch   **3**

3   No eye patch; entirely aquatic; body flattened; a
    powerful swimmer with large webs on hind feet
                African clawed frog, *Xenopus laevis* (Fig. 8.27, Map 10)

    Clawed frogs are very rare but also secretive, known for certain
    now only in a south Wales valley.

-   Not like this; little or no eye patch; usually in or near
    water all year round; (upper body colour often (but not
    always) at least partly green, often bright, sometimes
    with spots or a vertebral stripe; males call loudly in
    breeding season with lateral, pea-shaped vocal sacs)
    (Fig. 8.23). Water frogs   **4**

    Even experienced fieldworkers find identification difficult,
    perhaps unsurprisingly as edible frogs are hybrids of the other
    two species and there are overlaps in most diagnostic features.
    Spawn and larvae are indistinguishable. Often a designation
    as 'water frog' is the best possible.

Fig. 8.54 Identification of water frogs; T = tubercle (from Beebee and Griffiths, 2000). Reprinted with permission by Bas Teunis (artist) and from HarperCollins Publishers Ltd

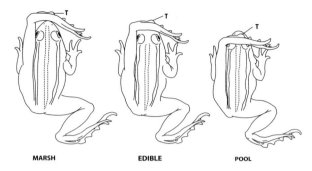

MARSH   EDIBLE   POOL

4   Ankle of hind leg extends to or beyond snout; metatarsal tubercle small (**Fig. 8.54**); up to 13 cm snout-to-vent length; (usually uniform green above, occasionally with pale vertebral stripe and/or dark spots; skin slightly warty; male vocal sacs grey)

Marsh frog, *Pelophylax ridibundus* (Fig. 8.25, Map 9)

-    Ankle of hind legs does not extend to snout   **5**

5   Ankle of hind leg extends to midway between eye and snout (**Fig. 8.54**); (medium size (up to 10 cm); male vocal sacs grey or white)

Edible frog, *Pelophylax kl. esculentus* (Fig. 8.24, Map 9)

-    Ankle of hind leg extends no further than the eyes; large metatarsal tubercle (**Fig. 8.54**); (relatively small (usually no bigger than 7 – 8 cm); male vocal sacs white)

Pool frog, *Pelophylax lessonae* (Fig. 8.23, Map 9)

If the upper body background is predominantly green in most specimens, and if it occurs in mixed populations with edible frogs, it is probably the introduced variety.

If the upper body is predominantly brown, and if it is not found with other water frog species, it may be the native variety.

6   Prominent parotoid glands behind the eyes; body up to 10 cm long   **7**

-    No parotoid glands; small (no bigger than 5 cm) (males sometimes with creamy-yellow eggs in strings wound around hind limbs)

Midwife toad, *Alytes obstetricans* (Fig. 8.26, Map 11)

Midwife toads are even less widespread than natterjacks and usually turn up in or near gardens.

7 No vertebral stripe; walks or hops feebly; does not run; up to 10 cm

> **Common toad,** *Bufo bufo* (Fig. 8.21, Map 7)

- Almost always with yellow vertebral stripe; runs rather than hops; (rarely bigger than 8 cm; males call loudly in the breeding season and have a single large vocal sac)

> **Natterjack toad,** *Bufo calamita* (Fig. 8.22, Map 8)

Natterjack toads have a restricted distribution and usually occur on sandy soils but common toads can also live in the same habitats. A toad found on sand is not necessarily a natterjack, indeed it often won't be.

Key IV Frog and toad spawn

1	Eggs in clumps of jelly	2
-	Eggs in gelatinous strings	4

2 Single large clumps, often clustered together to form a mat in shallow water; individual eggs black with white 'polar' spot

> **Common frog,** *Rana temporaria* (Fig. 8.28)

- Multiple small clumps or individual eggs   **3**

3 Eggs predominantly white with brown or black patches; (in small clumps or as individual eggs; scattered on pond bottom, not in shallow water)

> **African clawed frog,** *Xenopus laevis*

- Individual eggs brown with white polar spot; (in multiple small clumps; often hidden in weed in deep water but near the surface)

> **Pool, edible or marsh frog** (Fig. 8.30)

4 Eggs in a double row (sometimes triple or more when freshly laid); strings intertwined together and around submerged vegetation; normally in relatively deep water

> **Common toad,** *Bufo bufo* (Fig. 8.29)

- Eggs in a single row (sometimes double when freshly laid); strings normally isolated from each other; in shallow water (just a few cm deep), often trailing across the pond floor

> **Natterjack toad,** *Bufo calamita* (Fig. 8.31)

Key V Frog and toad larvae

1 Looks like a small catfish with a pair of front 'feelers';
swims in open water
African clawed frog, *Xenopus laevis* (Fig. 8.37)

- Not like this **2**

2 Upper body predominantly brown, sometimes spotted
or speckled **3**

- Upper body black **5**

3 Up to 3.5 cm long (snout to tail tip) when fully grown;
speckled with gold
Common frog, *Rana temporaria* (Fig. 8.32)

- Larger, up to 7 cm when fully grown; not speckled
with gold **4**

Fig. 8.55 Ventral
surface of common
toad *Bufo bufo* (A) and
natterjack toad *Bufo
calamita* (B)

4 Often with dark spots or blotches, especially on tail
fins; (large when fully grown, up to 7 cm, longer than
adults!)
Midwife toad, *Alytes obstetricans* (Fig. 8.36)

- Usually unspotted above; (sometimes with pink or
mauve tinge on belly; grows very large, up to 7 cm)
Pool, edible or marsh frog (Fig. 8.33)

5 Up to 3 cm long; (uniform black colour, **Fig. 8.55A**; often
shoal in open water)
Common toad, *Bufo bufo* (Fig. 8.34)

- Up to 2.5 cm long; (black, sometimes with white chin
patch, **Fig. 8.55B**); may congregate around pond edge,
especially in sunny weather
Natterjack toad, *Bufo calamita* (Fig. 8.35)

Although the larvae of common and natterjack toads are hard
to distinguish morphologically there are usually other clues.
Natterjacks spawn later than common toads and small tadpoles
in May or June, in shallow sandy ponds, are quite likely to
be natterjack. Clustering/shoaling behaviour also differs as
mentioned in the key. White chin patches are usually prominent
in well-grown natterjack tadpoles from Irish Sea coastal
populations but may be absent altogether in tadpoles from
southern and eastern England, so lack of a chin patch does not
necessarily imply that the individual is *B. bufo*.

8.3 Reptiles

Again there are two biologically distinct groups in Britain, notably lizards and snakes, but in this case the distinction is not immediately obvious in all cases. One of the commonest lizards in Britain is the slow-worm which is serpentine in form and frequently mistaken for a snake. It is therefore important to make this distinction before proceeding to a more general lizard key. Slow-worms vary in colour from grey through brown to gold or copper, generally uniform in males (though sometimes with small blue or brown spots) but often with a thin, black vertebral stripe and dark flanks in females. The stripe is also present in all young slow-worms (**Fig. 8.43**). No snakes have these colour patterns (see below). Slow-worms are also smaller than adult snakes, rarely exceeding 40 cm long, and have relatively short, blunt tails.

One of our native species (the sand lizard) and two introductions (green and wall lizards) lay eggs but these are always buried in sand or hidden deep in vegetation so are almost never encountered by naturalists. The following key covers adults of four species of limbed lizards and distinguishing features of juveniles in their year of birth. Later on they resemble small adults.

Sexing lizards

Male lizards tend to have brighter colours or patterns than females, broader heads relative to their bodies and a distinctive bulge at the base of the tail. Male viviparous lizards often have bright orange bellies while those of females are paler and more yellow. Male sand lizards develop bright green flanks in spring, but this colour intensity declines through the summer and by autumn has virtually disappeared. Male green lizards have a distinctly bluish throat in spring but usually no back stripes. The converse is true of females for both these characters. Females never develop blue throats and often have brown stripes along their backs.

Key VI Adult and hatchling limbed lizards

1	Larger than 6 cm	2
–	Smaller, less than 6 cm including tail	5

2	Up to 18 cm including tail, but usually smaller; back pattern mottled and/or striped; background may be greenish but is more often brown	3
-	Larger, adults commonly 20 cm and may be up to 40 cm; often with bright green on upper body or, if not, with distinct 'eye spots' on a brown or grey background	4

3   Rarely longer than 15 cm; back marked with flecks and stripes; (not often found on walls)
Viviparous or common lizard, *Zootoca vivipara*
(Fig. 8.39, Map 13)

-   Similar to viviparous lizard but slightly larger (up to 18 cm) and back pattern distinctive; (very agile; often found high on walls)
Wall lizard, *Podarcis muralis* (Fig. 8.41, Map 15)

Wall lizards are, as their name suggests, usually associated with walls of both ancient and modern human dwellings whereas viviparous lizards are more at home on open banks, forest rides and so on as well as on heathlands, moors and dunes.

4   Up to 20 cm long; stocky and with 'eye spots' on back; (may have bright green flanks)
Sand lizard, *Lacerta agilis* (Fig. 8.40, Map 14)

Sand lizards are confined to heathland and sand dunes or closely associated habitats.

-   Up to 40 cm long; predominantly green all over upper body including head, with fine black spots or pale/brown lateral stripes along back
Green lizard, *Lacerta bilineata* (Fig. 8.42, Map 16)

Green lizards are, at the time of writing (2011) known only from cliffs near Bournemouth.

5   Jet black when born (around 4 cm), almost uniform copper but with fine spots when a little larger
Viviparous or common lizard, *Zootoca vivipara*
(Fig. 8.44, Map 13)

-   Not like this   6

6 Grey-brown, with tiny 'eye spots' on flanks; about 5–6 cm at birth

> **Sand lizard, *Lacerta agilis* (Fig. 8.45, Map 14)**

- No eye spots   **7**

7 About 5–6 cm at birth; back brown with distinctive, pale dorsolateral stripes; may have purple sheen

> **Wall lizard, *Podarcis muralis* (Fig. 8.46, Map 15)**

- 7–8 cm at birth (pl. 3.4); light brown, with a hint of green around the mouth; two to four pale, maybe faint, dorsolateral stripes

> **Green lizard, *Lacerta bilineata* (Fig. 8.47, Map 16)**

8.4 Snakes

There are two species of snakes in Britain that lay eggs, the native grass snake and the introduced Aesculapian snake. Both seek artificial warmth in the form of piles of rotting vegetation, manure or compost heaps wherever these are available. Natural sites are also used but are less likely to be discovered by naturalists. Clutches of white, 2–3 cm-long leathery eggs found in compost heaps will almost certainly be those of grass snakes. Those of the much rarer Aesculapean snake are larger, up to 5–6 cm long and grooved. Juvenile snakes in almost all cases resemble adults so no separate key is provided. However, hatchling Aesculapian snakes strongly resemble grass snakes (both have creamy-yellow neck collars) but are larger at hatch, typically more than 20 cm compared with 15–18 cm for grass snakes.

Any snake found in Scotland, except possibly in the very far south, will almost certainly be an adder (**Map 17**). Any snake seen swimming will almost certainly be a grass snake. Aesculapian snakes are particularly fond of climbing quite high up in bushes or trees, although grass snakes also sometimes do this.

Sexing snakes

This is notoriously difficult for most species but adders are an exception. Males have pale, often silvery background colours with well demarcated black zig-zags while females are usually more brown or yellow with darker brown zig-zags. In all species the length of the tail, relative to body length, is greater in males than females, but assessing this requires capture and careful measurement. Large grass snakes (more than 90 cm) are virtually certain to be females.

Key VII Adult snakes

1   Body uniformly black (probably the dark form of)

Adder *Vipera berus* (Fig. 8.49, Map 17)

Any black or almost black snake will almost certainly be an adder. This colour variety is locally common.

\-   Body not all black   2

2   Background colour brown or green, uniform or with black vertical bars along flanks; (may have a yellow collar; often longer than 60 cm)   3

\-   Background colour grey, silvery or reddish-yellow, with brown or black markings along body; (almost always less than 60 cm long)   4

3   With a yellow collar; body usually with black vertical bars along sides. Up to 90 cm, sometimes larger

Grass snake, *Natrix natrix* (Fig. 8.48, Map 18)

\-   No yellow collar; body uniformly brown or green; (up to 150 cm)

Aesculapian snake, *Zamenis longissimus* (Fig. 8.51, Map 20)

Aesculapian snakes have only been confirmed as breeding populations at two localities, one in north Wales and the other in north London.

4   Background colour usually silvery or reddish yellow, with distinctive brown or black zig-zag markings all along body; occasionally overall dark or even black with no visible markings

Adder or viper, *Vipera berus* (Fig. 8.49, Map 17)

\-   Background colour grey or brown with dark irregular spots, not a zig-zag stripe, along the back

Smooth snake, *Coronella austriaca* (Fig. 8.50, Map 19)

Smooth snakes are very rare and confined to sandy heathlands in southern England.

8.5 Distribution of amphibians and reptiles

Approximate distributions of all the species listed above, at the national scale, are shown in the maps.

Local distributions

Various factors influence the distribution of amphibians and reptiles at local scales and this information, mostly concerned with habitat preferences, is useful when searching for them. A summary of this advice is given in **Table 8.1**.

Further information about identification and distributions of amphibians and reptiles in Britain and mainland Europe is available in several recent texts (Griffiths, 1996; Beebee and Griffiths, 2000; Arnold and Ovenden, 2002).

Table 8.1 Where to look for amphibians and reptiles. Widespread favoured habitats are listed but these are not exhaustive. For amphibians it is important that aquatic and terrestrial habitats are close together, ideally integrated into a common landscape. Small pond: less than 10 square metres; medium pond: 10–100 square metres; large pond: more than 100 square metres.

Species	Terrestrial habitat	Aquatic habitat
Amphibians: newts		
Smooth newt	Lightly grazed pasture, scrub, open woodland, gardens	Small or medium-sized vegetated ponds, pH > 5.5, with few or no fish
Palmate newt	Lightly grazed pasture, scrub, open woodland, gardens, heaths, moors	Small or medium-sized vegetated ponds, pH > 4, with few or no fish
Great crested/Italian crested newt	Lightly grazed pasture, scrub, open woodland	Medium-sized vegetated ponds, pH > 5.5, with no fish
Alpine newt	Gardens, parks	Small or medium-sized vegetated ponds, pH > 5.5, quite often with fish
Amphibians: frogs and toads		
Common frog	Lightly grazed pasture, scrub, open woodland, gardens, moors	Small or medium-sized vegetated ponds, pH > 5.5, often shallow and with few or no fish
Common toad	Lightly grazed pasture, scrub, open woodland, gardens, moors	Medium or large ponds, can be poorly vegetated, pH > 5.5, often with fish
Natterjack toad	Sand dunes, lowland heaths, upper saltmarshes	Small or medium-sized shallow, usually temporary ponds, pH > 5.5.
Clawed frog African clawed frog		Small or medium ponds of moderate depth, not necessarily vegetated
Water frogs	Low-lying pasture, open fields and commons, gardens	Medium or large ponds or ditches, well-vegetated, pH > 5.5, often with fish
Midwife toad	Gardens	Garden ponds, usually without fish
Reptiles: lizards		
Slow-worm	Scrubby hillsides, open woodland, gardens, heaths and moors	
Viviparous lizard	Scrubby hillsides, open woodland, heaths, dunes and moors	
Sand lizard	Sandy lowland heaths and coastal dunes	
Wall lizard	Around buildings (modern and ancient) well exposed to the sun; also cliffs and scree banks	
Green lizard	Scrubby south-facing cliffs	
Reptiles: snakes		
Adder	Scrubby hillsides, open woodland, heaths, dunes and moors	
Grass snake	Lightly grazed pasture, scrubby hillsides, open woodland, heaths, dunes and moors	Small or medium-sized ponds and ditches well used by amphibians
Smooth snake	Sandy lowland heaths	
Aesculapian snake	Areas with dense vegetation, in or near zoological gardens	

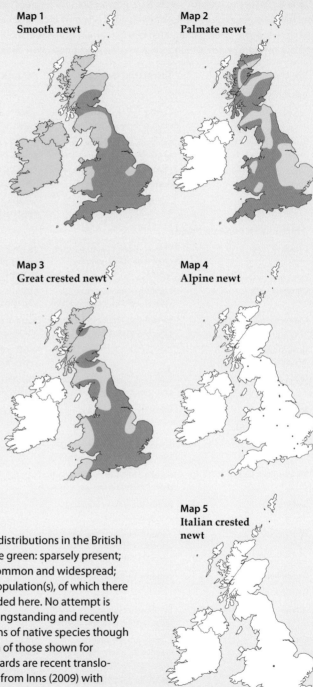

Map 1
Smooth newt

Map 2
Palmate newt

Map 3
Great crested newt

Map 4
Alpine newt

Map 5
Italian crested newt

Maps 1–20
Amphibian and reptile distributions in the British Isles. White: absent; pale green: sparsely present; dark green: relatively common and widespread; purple: reintroduced population(s), of which there may be more not recorded here. No attempt is made to differentiate longstanding and recently translocated populations of native species though a significant proportion of those shown for natterjacks and sand lizards are recent translocations. Mostly derived from Inns (2009) with permission from WildGuides Ltd.

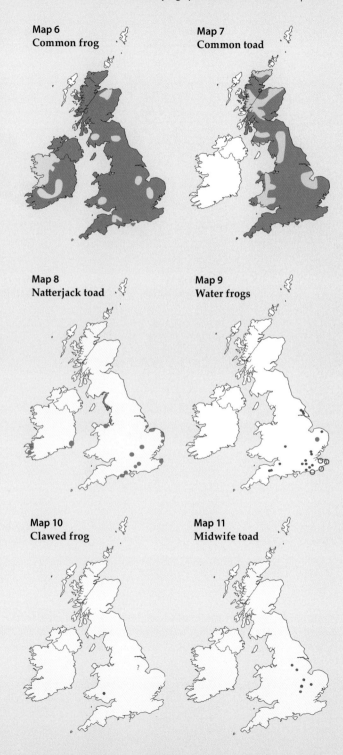

Map 6
Common frog

Map 7
Common toad

Map 8
Natterjack toad

Map 9
Water frogs

Map 10
Clawed frog

Map 11
Midwife toad

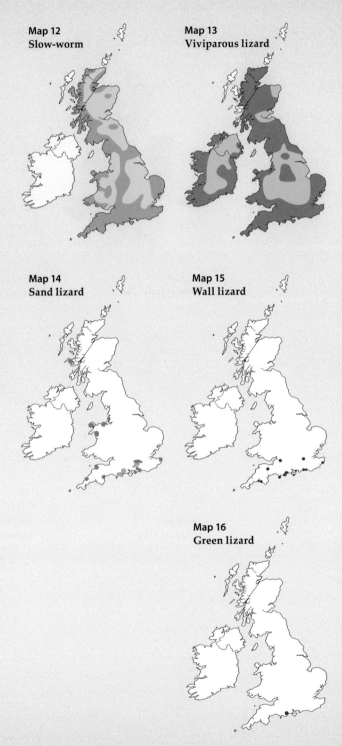

Map 12
Slow-worm

Map 13
Viviparous lizard

Map 14
Sand lizard

Map 15
Wall lizard

Map 16
Green lizard

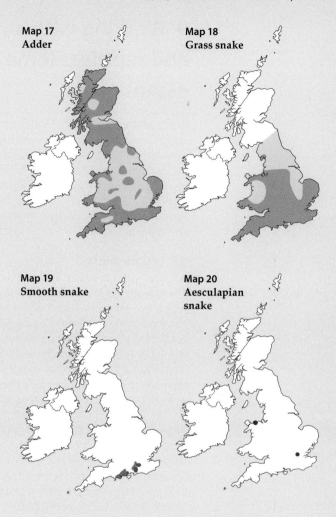

Map 17
Adder

Map 18
Grass snake

Map 19
Smooth snake

Map 20
Aesculapian snake

9 Working with amphibians and reptiles: some basic essentials

Before getting involved in detailed proposals for research on amphibians and reptiles it's as well to consider some general aspects of working with these animals. In this chapter issues are discussed which impact on legal, practical (handling, identification of recaptured individuals) and statistical aspects. This discussion is designed to help in the planning of any investigation on the natural history of amphibians and reptiles.

9.1 Legal niceties

Before embarking on any work with amphibians and reptiles it is important to understand the legal framework within which everyone must operate for identifying, surveying and monitoring wild animals (chapter 4) as well as for more detailed studies of the kind suggested in chapters 5–7. The law is not especially onerous and for many kinds of work it has little or nothing to say. However it is necessary to be aware of three particular areas of legislation, in increasing order of relevance to home-based and field studies.

The Dangerous Wild Animals Act (1976) restricts the keeping of species considered a possible danger to the public. The adder, controversially in the opinion of many herpetologists, is included under this Act. This means it is necessary to obtain a licence from your local authority to keep adders in captivity. You will be charged for the privilege and will have to satisfy inspectors that any enclosure constraining adders is both escape-proof and secure from public entry. None of this applies to working with adders in the field.

The Animals (Scientific Procedures) Act (1986) limits what you are allowed to do with respect to invasive techniques and vertebrates. The law was framed primarily to provide safeguards for animals in laboratories but it applies to field work as well and potentially also to experiments undertaken at home. Licences under this Act are essentially unobtainable for amateurs; they are invariably awarded to institutions and thence to their individual employees. It is therefore essential for amateurs to avoid using procedures that would require a Home Office licence. This issue rarely arises in practice and is certainly not relevant to survey or monitoring studies. Handling amphibians and reptiles and keeping them in captivity are not 'procedures', and therefore do not need these licences. Even injecting animals with PIT tags no longer comes under Home Office constraints, though it did at one time. The kinds of things you cannot do without a licence include clipping off toe tips or conducting experiments to see which predators attack tadpoles, or adult amphibians and reptiles for that matter. Most of us wouldn't choose that kind of work anyway but ironically you can still do many of these things as a 'non-scientist', that is, not as part of a deliberate experiment that will eventually be written up and published. There are still 'grey' areas, and the types of investigations suggested in this book were chosen to ensure that they should not fall foul of this Act.

The most relevant pieces of legislation for amateurs and professionals working on amphibians and reptiles in Britain are the The Wildlife and Countryside Act (WCA, 1981) and its successors, including the Countryside and Rights of Way Act (CROW, 2000) in England and Wales, the Nature Conservation (Scotland) Act (2004), and the Wildlife (NI) Order (1985) in Northern Ireland. All of these laws relate to conservation rather than animal welfare and concern habitats as well as species. The requirements of these Acts are therefore described in detail below.

The various conservation Acts confer up to three possible protection levels on our native species. The widespread amphibians (common frogs and toads, smooth and palmate newts) in most of Britain receive the lowest level. No licences are needed for anything except collection for commercial sale. Common frogs in Northern Ireland and widespread reptiles in the rest of the UK (slow-worms, viviparous lizards, adders and grass snakes) enjoy partial protection; it is illegal to injure or kill them, as well as to sell them, without a licence. It is legal to survey, catch and handle them, and once again, therefore, these laws do not impinge at all on anyone wanting to study these animals. There is however a high level of protection applying to great crested newts, native pool frogs, natterjack toads, sand lizards and smooth snakes and, in Northern Ireland, to smooth newts and viviparous lizards. For these species licences issued by the statutory (government) conservation agencies are required. Even disturbing these animals during survey or monitoring (for example, using a high power torch or bottle traps, or looking for eggs when searching for great crested newts) requires a licence. The authorities expect some evidence of proficiency such as attendance at a training programme for the relevant methods before granting such licences. This is not a difficult hurdle, though, and anyone interested in working with the rarer species need not be deterred by what is usually a simple and reasonably quick qualification process. The licences themselves are free of charge once granted.

None of the introduced species are protected by law but the legislation is nevertheless relevant. It is illegal to introduce any non-native amphibians and reptiles into Britain. Furthermore, it is against the law to release any such animals, once caught, back into the wild even at their place of capture. Strictly speaking they should be held in captivity or killed. This inevitably causes problems for surveyors wishing to catch animals to confirm identification or to carry out any investigation that requires handling them. Whatever their legal status, most people would not want to kill and may not be in a position (or wish) to keep these creatures. Under some circumstances the statutory agencies will licence release back at the site of capture, but a bureaucratic process to approve this has to be endured. In practice it is often possible to identify our non-natives without catching them and this is generally the best option.

9.2 Handling amphibians and reptiles

A sensible philosophy is to avoid handling any animal unless it's really necessary, and then minimise the time doing it. Handling is potentially stressful for all species and in some cases can also be unpleasant for the people involved. Many studies require little or no handling. Reptile survey, for example, can be based purely on observation. But situations may arise where the animals must be caught for identification purposes (particularly for some amphibians), or for experimental studies. Disposable gloves can be an asset but use nitrile rather than latex ones; some amphibians react badly to latex.

Adult frogs, toads and newts are easy to catch and restrain by hand or using a sturdy pond net. Keep your skin or glove moist while you hold them, though, and wash afterwards. Many amphibians produce smelly or mildly toxic secretions and though none of the British species are remotely dangerous their output won't taste nice on sandwiches. Toads in captivity become tame enough to accept handling without obvious stress, but wild-caught animals of all species are obviously traumatised and will try to escape. Toads often empty their bladder when first caught but it's some consolation that this is a water reserve, not urine. Nevertheless toads need such a reserve, especially in dry weather or arid habitats, so handling could conceivably threaten survival if the reserve can't be quickly replenished. Frogs, being so agile, are particularly difficult and handling should be kept to an absolute minimum. Catching water frogs is especially problematic. It is reasonably easy at night, using a torch and pond net, but in daytime they dive rapidly out of sight and are hard to take unawares. The animals should be released at or near the place of capture as soon as possible (but remember the rules about introduced species), unless they are needed for study at home. Always transport adult amphibians in damp conditions (maybe in a plastic sandwich box or bucket) with a bit of vegetation but not in water, and not exposed to the sun.

Larvae should virtually never be handled since they require submersion under water to breathe. They should be kept submerged and examined in a glass or clear plastic-walled container. Some larvae, especially those of great crested newts, are rather fragile and easily damaged by rough netting during capture. Take care to be as gentle as possible.

Limbed lizards (as opposed to slow-worms) are extremely agile, and it takes skill and patience to catch them. It's sometimes possible to creep up close and simply grab a lizard by hand. With practice this can work well, and it's the only option for juvenile animals too small to catch any other way. It works better for some species (sand lizards, which are relatively slow) than others (wall lizards, which are extremely quick and vigilant). There is a risk of damaging the lizard if you press down too hard or of inducing it to shed its tail if you land on that bit. This is a bad result because losing a tail amounts to losing an important predator defence mechanism. It grows back but this is a slow process and the regenerated tail is never as complete as the original. The animal is also sacrificing a food reserve as nutrients are stored in tail tissues. An altogether better method that also requires skill and practice is noosing. A thin nylon noose (low breaking-strain fishing line is good)

Fig. 9.1 Noose for catching lizards

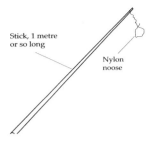

Stick, 1 metre
or so long

Nylon
noose

with a loose knot is attached to the end of a stick (**Fig. 9.1**). Lizards generally allow a stick to approach them closely without panicking, provided movements are slow and deliberate. The noose is gently dropped over the head as far back as the neck and then lifted quickly to tighten it, hopefully with a struggling lizard attached. This should be transferred immediately to your other hand. Hold the animal gently but firmly (not near the tail!), relax the noose and remove it as fast as you can. Lizards are not physically damaged by this procedure. Often they don't see the noose at all until it's too late, but if they do it's sometimes mistaken for food, presumably a fluttering insect. It's amusing but frustrating to watch the lizard you are trying to catch, itself trying to catch your trap. Lizards may bite when caught but no British species can draw blood – it's only a nip. If you need to take the lizard somewhere, cloth bags tied with a cord around the top are ideal for transport. As with amphibians, make sure they are not left in sunshine to overheat.

Slow-worms and snakes are normally caught by hand. They are not as fast as lizards but are quick enough to escape unless you pounce rapidly. With slow-worms, again avoid the tail because it's shed just as readily as those of the limbed lizards. Slow-worms and grass snakes may thrash about when picked up although grass snakes sometimes feign death and both species are likely to discharge the foul-smelling contents of their cloacas over you. Be prepared! Neither, however, is likely to bite. Smooth snakes on the other hand quite often chew on a finger, occasionally after a minute or two of apparent docility, thus producing quite a surprise. Although they can draw blood there's no poison and little pain, but disinfect the bite as soon as you can. Adders of course are a different matter. If you must pick an adder up the best plan is to grab the tail close to the end as it moves away from you and lift it up sharply, well away from your body. Adult adders are too heavy to rear back up and bite your hand if they are held so they hang vertically downwards but they might swing out and catch any nearby body parts! Don't try this with small adders which are supple enough to bend upwards and bite. The old fashioned ideas about using forked sticks to pin down venomous snakes behind the head are pretty useless and can damage the snake so don't try that. But the main point is not to catch adders at all unless you have a very good reason and then only do so with guidance from an expert. Like lizards, snakes can be transported in large cloth bags tied around the bag's neck.

9.3 General good practice

It goes without saying that permission of the landowners must be obtained before starting any work in the field. Usually this is not a problem and many owners are very interested in, and proud of, the wildlife on their patch. Safety considerations are also important. Ponds in particular can be dangerous places with some risk of drowning if care is not taken. Always check out a proposed survey site in daytime to find out where potential danger might lurk (steep banks, potholes) and ideally, especially at night, work with at least one other person. Take special care checking out hiding places where there might be adders. A bite is unlikely to be serious but it certainly won't be a pleasant experience. Pond water in Britain is not a significant health problem but Weil's disease, caused by bacteria excreted in rat urine, is on very rare occasions contracted by exposure (usually via open cuts) to pond or river water. Don't eat lunch while out netting unless you have sterile wipes or clean water for hand washing! Keep a charged mobile phone on hand for emergencies. Surveying for amphibians and reptiles is far from a hazardous occupation and I'm not aware of any serious incidents but it's clearly best to minimise risks.

An issue of concern recently has been cross con-tamination of ponds by non-native species, and this needs consideration at the start of an amphibian survey. Invasive plants such as New Zealand stonecrop (*Crassula helmsii*) and parrot's feather (*Myriophyllum aquaticum*), among others, are now widespread in Britain and when established in a pond can out-compete native species. A tiny fragment carried inadvertently to a new pond is enough to precipitate an invasion by vegetative growth. It is therefore good policy to make sure that equipment (boots, nets, newt traps) is checked and cleaned if necessary between visits to different ponds. This may also help reduce the spread of diseases such as *Ranavirus* and chytrid fungus.

9.4 Individual identification

For some types of study it is useful to be able to recognise the same individual when caught on different occasions. In the past this was done using a wide range of techniques, most of which are now redundant or not legal for amateurs. However, for short-term marking a couple of old methods are worth considering. For studying a small population of frogs or toads in a garden pond, individuals can be fitted with waist bands attached to a coloured or numbered tag. The bands should be made from material that will disintegrate naturally within a few weeks, such as the absorbable sutures used in surgery, and thus will not stay with the animals for long after they leave the pond. They must be loose enough not to cause skin damage but tight enough so that the frogs can't wriggle out of them; this skill is learnt by experience. An advantage is that tags can be identified by observation from a distance with no need for re-catching and handling. Some of the classic early studies on frog breeding behaviour used this method very successfully but it's not much help for newts or reptiles. However another method can be used with lizards or even snakes: applying small spots of non-toxic paint, nail varnish or Tippex. The marks will only last until the next skin shedding, probably no more than a few weeks, but again can allow identification without re-catching. None of these materials should ever be applied to amphibian skin which is permeable and easily damaged.

Most research requiring individual identification now employs PIT tags (**Fig. 9.2**) or digital photography. Both are possible options for amateurs but photography is undoubtedly the cheapest and most flexible in terms of the number of species and life stages it is appropriate for. Indeed, it would be valuable to investigate the range of species for which it is suitable more fully than has yet been done and try to include potentially 'difficult' species. Smooth and palmate newts and viviparous lizards have spot patterns and other markings but are they variable enough to generate reliable individual 'fingerprints'? A practical consideration for photographic identification concerns the requirement to take pictures from a consistent angle. This makes for easier cross reference in a set of individual photos as a collection builds up. In some cases, as with basking lizards, it may be possible to obtain adequate pictures without catching the animals but this is not a common situation. Usually the individual is constrained within a fairly tight-fitting clear plastic tube while the photograph is taken. In small populations previously caught individuals can be recognised by visual inspection of the picture library.

Fig. 9.2 Passive Integrated Transponders (PIT) tags allow unique alphanumeric codes to be assigned to individuals for mark/ recapture experiments. (Tag: Light Warrior; reader: USFWS Mountain-Prairie)

However, software exists to screen large collections and is increasingly tolerant of variations in camera angle. Programs for great crested newts, sand lizards, adders and frogs were obtainable (at a price) at the time of writing from www.conservationresearch.co.uk. Freeware has recently become available as 'Wild-ID' from http://software.dartmouth.edu/Macintosh/Academic/Wild-ID_1.0.0.zip and more is under development at the University of Kent ('Zoometrics', www.eda.kent.ac.uk/research/theme_project.aspx?pid=57).

9.5 Forward planning and the basics of statistical approaches

A vital step is to plan any proposed study carefully and decide what you want to do with the results. You will probably want to make the outcome available to other people by publishing it in the public domain. Rigour of project design is particularly important for a full-blown investigation aimed at a scientific journal. It's worth stressing that valuable and interesting contributions do not necessarily require the discipline of strict scientific planning. One-off observations and other novelties are, for example, regularly published in the *Herpetological Bulletin* (see chapter 10) under 'natural history notes'. On the other hand if a rigorous investigation is planned either in the field or under captive conditions, statistical analysis of the results will be an essential prerequisite for submission to a peer-reviewed journal (see **9.7** for suggestions). Furthermore it is important to consider statistical aspects at the design stage of the study. It can be very disappointing if, at the end of the work, it turns out that the design was not amenable to rigorous analysis. Some guidelines about statistical methods and their application are given below.

Hypothesis testing

The fundamental objective of standard statistical tests is to assess the likelihood, or probability, that a particular hypothesis made at the start of an investigation is true. A simple example of such a hypothesis is:

'The newts in pond A are significantly smaller than the newts in pond B.'

'Significantly' is a key word here, meaning that any difference is probably real and not merely a chance result caused by inadvertently biased sampling. Any hypothesis is essentially an idea based on some incidental observations or questions that have arisen in your mind. Statistics will help to convince you, and then other people, whether or not your idea is valid. Paradoxically the starting point is normally a 'null hypothesis', that is, the hypothesis that your idea is not true! In the example above this would be that there is actually no difference in the sizes of newts in the two ponds. The statistics then yield a probability (ranging from 0 = impossible to 1 = certain) that the null hypothesis is correct. If this probability (P) turns out to be 0.05 or less, meaning that there is no more than one chance in twenty that the two sets of newts are identical, this is conventionally accepted as indicating that the null hypothesis can be rejected; there is a 'significant' difference and your idea is supported. The lower the P value, the stronger is the support. It's important to realise that the 0.05 value of P is arbitrary. Is a result of $P = 0.06$ really so different and not at all significant? The relationship of P values to 'significance' makes more sense in the context of 'type 1' and 'type 2' errors. With P set at 0.05 it follows that on average, one time in twenty when this value crops up in a series of studies the conclusion of significance will be wrong by chance and the null hypothesis actually correct. This is a 'type 1' error – rejecting the null hypothesis when it ought to be accepted. The lower the observed P value, the less likely is an error of this kind. The converse error, in most cases less of a worry, is to accept the null hypothesis when it should be ruled out, thus generating a 'type 2' error. This is where P values of slightly greater than 0.05 become problematic. Type 1 errors are the most important because they can support a false hypothesis and the emphasis should normally be on reducing them rather than type 2 errors. Always aim for a research design that is likely to yield results with the highest possible level of significance. In practice this worthy ambition is usually constrained by operational practicalities such as the number of replicates that can be accommodated within reasonable resource limits.

One more thing. If you carry out several independent but similar tests this must be accounted for when assessing probabilities. Let's say you compare the relative sizes of two species of newts in the same two ponds. Then you will end up with two test results, one for each species. But now there is a 1/20 plus another 1/20 (= 1/10) chance of getting a 'significant' result just by chance in at least one of the two comparisons. To accommodate this it is normal to carry out a 'Bonferroni correction' by dividing 0.05 (the standard threshold P value for significance) by the number of independent tests, in this case two. So you now need P values as low as 0.05/2, = 0.025 in at least one of the tests to be convinced of significance if the other test yields a P value greater than 0.05. On the other hand if both tests yield P <0.05 there is less cause for concern about a spurious result. In general for multiple comparisons use the sequential Bonferroni procedure. Thus in three comparisons the most significant should have P<0.05/3, = 0.017; the second should have P<0.05/2, = 0.025; and the third is significant if P<0.05.

There is a large and increasing array of statistical tests available for ecologists and they are still multiplying on account of ongoing increases in computer power that permit complex calculations that were utterly impossible a decade or two ago. Fortunately longstanding and relatively simple tests suffice for most purposes. Useful texts that give user-friendly instructions about specific tests, more comprehensive than there is space for here, include those of McGarigal, Cushman and Stafford (2002) and Dytham (2010). Wheater and Cook (2003) provide an excellent guide to data analysis, including statistical methods appropriate for ecological studies. Software packages are required to perform all but the simplest of tests. Most of those used in academia are under site licences, among the most popular of which are MINITAB and SPSS, but they are not cheap to buy. However, EXCEL can carry out many of the simpler analyses and there is free downloadable software from the internet that includes most of the tests you are likely to need (see for example http://en.wikipedia.org/wiki/Free_statistical_software). Failing that, calculations for some of the most straightforward analyses can be done with a pocket calculator using formulae provided in the various textbooks or on line. In this case when you have calculated the test statistic you will need to look at tables (also provided in textbooks and on line) to determine the probability and thus significance of your result. At this stage when consulting the tables you need to take account of 'degrees of freedom'. This number is often but not always one less than the sample size, so if you had 20 newt measurements from each pond, in the simplest tests there would be 38 degrees of freedom. Most of the software packages do all this for you.

9.6 Statistical methods

The following sections outline procedures applicable to the kinds of study proposed in other chapters, starting with a basic strategy that must be considered at the outset. There are then a few examples of some of the commonest tests, mostly without mathematical details. This chapter is therefore not intended as a substitute for a thorough statistical text book, merely as a general guide of approaches that can be taken.

Planning the study

The first thing to establish is whether your project plan has a large enough sample size or enough replication to show significant differences if they truly exist. Random problems can obscure what is really going on if you only have one example of each treatment and it is impossible to overemphasise the need to replicate trials in ecological experiments. Assessing this requirement is referred to as *a priori* power analysis. Ideally you want to start off with at least an 80% chance that your design will be up to the job. *Post hoc* power analysis after the results are in is also possible but isn't much use if you discover too late that the design was poor. Fortunately there is free software for *a priori* analysis (G*Power, available via www.psycho. uni-duesseldorf.de/abteilungen/aap/gpower3) to help with this task. Even so, you have to make a subjective judgement at the start about 'effect size' – anticipate whether the differences between samples are likely to be 'small', 'moderate' or 'large'. Logically enough, you will need a larger sample to establish the significance of a small difference than a big one. Back to the newt example: if you expect the differences between ponds to be small, power analysis indicates you will need 620 individuals from each pond to have an 80% chance of proving your point. On the other hand if the variation seems large you will only need 42 newts per pond (all using a 't-test', see later). These estimates are based on so-called 'one tailed' tests where you predict the direction of difference (that newts in pond A are smaller than those in pond B). If you merely predict a difference but not the direction of it (newts in A and B aren't the same but without anticipating which are bigger) then a 'two-tailed' test is appropriate and this needs even larger sample sizes for the equivalent power. It is of course possible to accept a lower than 80% chance of success. So if you're happy with 70%, for example, the sample size requirement for a large effect drops to 32 newts per pond.

Finally, a warning about 'pseudoreplication'. If you run an experiment to measure tadpole growth rates you might have three aquaria for each test condition (perhaps for each type of food), and five tadpoles in each aquarium. How many replicates does this represent? The answer is three, not 15, because within each aquarium the tadpoles cannot be considered independent. They might all be influenced by a common factor such as the exact position of the aquarium. So the correct procedure in this example would be to use the average growth of each set of five tadpoles as the individual data points (giving three data points for each treatment). Always be conscious of this need for independence in data. Similar considerations apply, for example, if you test different habitats by putting down groups of refugia in different places and checking each group several (say, five) times through the spring. The number of, say, adders to use in any subsequent analysis should be the average seen under each group of refugia, averaged over the five visits, and not the five individual counts. So for each site there would be one number for each group of refugia, assuming that sequential use of each is not independent (because the same animals often return many times) but individual groups of refugia across the site are more likely to be independent.

Investigating variations in distribution

Is the distribution of an animal random in its environment or is it affected by different habitat structures? For example, maybe you catch 100 newts altogether in a pond but far more in some regions than in others. A hypothetical example is shown in **Table 9.1**. Assuming roughly similar amounts of each habitat type (not very realistic, but variations can be compensated for), it's easy to predict how many newts you would expect in each habitat if they are not choosy, which is the null hypothesis of no habitat selection. You then test whether the observed numbers differ from the null expectation using a χ^2 (chi-squared) test.

Table 9.1 Hypothetical newt distributions with a total sample of 100

	Open water	Dense vegetation	Muddy shallows	Reedbed	Duckweed cover
Observed number	30	50	5	10	5
Expected number if random	20	20	20	20	20

The χ^2 statistic is simply calculated:

$$\chi^2 = \sum \frac{(O - E)^2}{E}$$

Where O = the observed number and E = the expected number. In the above example, then,

$$\chi^2 = \frac{(30 - 20)^2}{20} + \frac{(50 - 20)^2}{20} + \frac{(5 - 20)^2}{20} + \frac{(10 - 20)^2}{20} + \frac{(5 - 20)^2}{20}$$

which = 77.5. Consulting χ^2 tables with n-1 (n = five habitats) = four degrees of freedom, the probability of the null hypothesis being correct is less than 0.01. Clearly the newts were not randomly distributed and preferred dense vegetation or open water to the other available niches. χ^2 calculations are often referred to as 'goodness of fit' tests, comparing data with theoretically predicted frequencies. They should never be used, however, in situations where any of the expected numbers are below five. In such cases it is often possible to pool data (for example, from two marginally used habitats) so that together they have an expected number of five or more.

Data distributions

Many statistical procedures are all about looking at differences between or among populations, as in the example of newt sizes outlined earlier, looking for correlations between population size and some other variable (perhaps between newt numbers and pond size) or explicitly trying to explain observations (again, maybe newt numbers per pond) by regression against variables that might include pond size, vegetation cover, pH and so on. Before choosing a specific method for any of these tasks, the first job is a technical one: assessing how your data are distributed.

The most powerful tests for all of the above jobs assume that the data are 'normally distributed', that is, when plotted as frequency classes they generate the familiar 'bell-shaped curve'. If this is true, then 'parametric tests' can and should be used. Statistical software packages provide methods for checking this out, such as the Shapiro-Wilk test. If the data are not normally distributed it is sometimes possible to make them so by using transformations. Common transformations include converting data into $\log_{10}$ or square root values, then checking again to see whether they now fulfil normal expectations. If this fails you will need to apply the corresponding 'non-parametric' tests. When unsure or when sample sizes are small (fewer than ten) it is safest to use non-parametric tests anyway. Some types of data are not expected to comply with normal distributions and need more specific analytical procedures. One such situation relates to survivorship studies. An animal is either dead or alive so results of this kind inherently have a binomial distribution with just two possible states, rather than a normal distribution.

Comparing means

Are those newts in pond A really smaller than those in pond B?

For normally distributed data the commonest procedure for comparison is the t-test, which generates a t-statistic and its associated probability. For this and other parametric tests comparing means (such as ANOVAs, see below) it is also important to establish that both data sets have comparable variances, defined as s^2 where s is the 'standard deviation' of the sample (note, of the sample, not of the population). This in turn is estimated as:

$$s = \sqrt{\frac{\sum x^2 - \frac{(\sum x)^2}{n}}{n - 1}}$$

where x = individual measurement and n = sample size. Again most software packages will compute whether variances of the two samples are sufficiently similar to justify the subsequent test. The analysis is tolerant of quite substantial differences in variance provided the means are also very different. A hypothetical example is given in **Table 9.2**.

Examining these data, both sets are normally distributed by the Shapiro-Wilk test and both have reasonably similar variances (7.66 mm for pond A, 8.46

Table 9.2 Some invented newt size measurements (in mm, snout to tail tip), with a total sample size of 20 individuals.

Pond A	Pond B
76	82
80	86
77	88
83	84
81	80
75	85
76	83
79	85
82	89
80	81

mm for pond B). However, the mean size of pond A newts (78.9 mm) is more than 5 mm smaller than the mean for pond B newts (84.3 mm). t-test analysis with (20 - 2 = 18) 18 degrees of freedom (subtracting one from each pond's sample size) indicates that this difference is highly significant with $P = 0.0005$. Because the data set is rather small, not enough to be really sure the data are normally distributed, it is worth trying a non-parametric equivalent (Mann-Whitney U test) as well. These tests make no assumptions about how the data are distributed but simply rank them in order of size across both samples. In this case the estimated significance is $P = 0.0019$; still strong evidence but not as strong (because the P value is higher) as indicated by the parametric test. This is a typical result with the parametric test giving higher significance (a lower probability) than its non-parametric equivalent.

In some cases another powerful method for comparing means is possible. This arises when the individuals from two samples can be paired in specific ways (so could not apply to the newt data as shown in **Table 9.2**). However, instead of pond A and pond B consider that the data columns in **Table 9.2** relate to the same set of 10 newts caught in the same pond in year 1 (A) and again in year 2 (B). You can identify the newts individually by their belly spot patterns, so you know you are comparing like with like. Now you can carry out a 'matched pairs' t-test (parametric) or a 'Mann-Whitney U test' (non-parametric) giving P values of 0.0018 and 0.0029 respectively. Yes, the newts have grown!

These examples produced significant differences even with sample sizes smaller than those suggested by power analysis. Indeed even with a large effect size the 'power' should only have been about 50% which all goes to show that although *a priori* power estimates are always desirable it is sensible to interpret them liberally. They need not necessarily preclude a study when the effect size may be very large as in this example.

A more complex situation arises if the comparison is not just between newts in two ponds but among samples from many ponds. For five ponds we would have five columns of data. In this case a one-way analysis of variance (ANOVA) is the appropriate method, again requiring roughly similar variances in each sample. And again there is a parametric version and a non-parametric equivalent (the Kruskal-Wallis one-way ANOVA). The results of these tests initially just tell you whether all the pond size profiles are indistinguishable (the null hypothesis) or whether one or more differ from the rest. 'Comparison of means' tests then follow to show specifically how each pond differs from the others. ANOVA is not limited to 'one- way'. It is possible to investigate more than one variable at the same time, perhaps 'sex' as well as 'pond' in the newt example. In a two-way ANOVA the influences on newt size of not only both these variables but also of any possible interactions between them can be analysed all at one go. In this case you might have 10 measurements of each sex from each of five ponds, so a total sample size of 100. If sex differences exist they may differ between ponds, implying some interaction between pond and sex that influences how big a newt will be. Perhaps, for example, both sexes are of similar size in pond A but females are bigger than males in pond B. ANOVA will reveal any such interesting facts, provided of course there is sufficient statistical power. And it would lead to an obvious follow-up question: why might this be?

Correlation and regression
Relating one variable to another is the basis for both correlation and regression. Is the size of a newt, for example, significantly related to its age? With correlation we simply investigate the strength of any such relationship which is measured using Pearson (r) or Spearman rank (r_s) correlation coefficients for parametric and non-parametric analyses respectively. In both cases possible

values range between 0 (no correlation) and + 1 or − 1, completely correlated either positively or negatively respectively. An associated probability is produced by the analysis. Regression takes this further, with normally distributed data, to produce an equation that precisely defines the relationship between the two variables. In this case another important statistic, R^2, is generated (again with associated probability of the regression's significance). R^2 is the proportion of variance in the dependent variable (newt size) that is explained by the independent one (newt age). A perfect linear relationship has an $R^2 = 1$. The null hypothesis for correlation and regression is lack of a relationship between the variables.

Before carrying out either of these analyses it is worth plotting the data to see whether a linear relationship looks likely. Even if there is a relationship, often it is not linear across the entire data range. For example growth often declines with age and after an initial linear effect 'flattens off' to generate an asymptotic curve. In these situations the simplest solutions are either to confine the analysis to the linear range or to attempt data transformations (see earlier section) to convert a curve to a straight line. In neither correlation nor regression is it safe to assume that significance implies cause and effect, although this simplistic conclusion is frequently drawn. It may be true but often warrants further study because the relationship might be coincidental and really produced by another, unknown factor. And which is really the dependent variable? Newt size could be a simple function of age if growth is continuous throughout life. But perhaps survival during the early growth phase is mainly determined by foraging success. The best hunters in this key period grow the fastest, have the lowest mortality rates and live to old age. In a sense, then, age is a function of size. Very often correlations and regressions require more investigation to disentangle these complexities. Even so, they are valuable steps in hypothesis testing provided the results are interpreted with due caution.

Survivorship studies

Survivorship data can be analysed like any other by applying non-parametric techniques of the types already mentioned. If numbers of surviving tadpoles, for example, are recorded at the end point of an experiment with five replicates of each of two treatments (maybe two different food provisions), a Mann-Whitney U test can be used to determine whether any survival differences are significant. However a more sophisticated approach investigates not just the end point but also any variation in survival rates over time. A range of tests taking account of binomial data distributions, mostly developed for clinical trials and other medical applications, is available to compare mortality rates under different treatments and some (such as the Gehan-Wilcoxon, Logrank and Peto-Wilcoxon tests) are available in common statistical software packages.

Multivariate statistics
It is in this complex area that statistical methods have expanded most dramatically, largely because of the possibilities opened up by ever increasing computing power. Multivariate analyses take many forms and only a few of the most relevant ones are mentioned here. Software packages are essential to implement any of them. Their common aim is to investigate the effects of many variables simultaneously and try to integrate their effects in a way that represents nature better than looking just at single variables, one at a time, could hope to do. In ecological field studies this is highly relevant because, for example, there may be many factors (agricultural practice, rainfall, temperature and so on) influencing the abundance of a plant or animal species. An important point about all multivariate techniques concerns the ratio of sample size to the number of variables considered. The former should always be larger, preferably much larger, than the latter and the temptation to include as many variables as possible should be resisted. The best way to ensure this is to always keep hypothesis testing in mind: focus on variables that you believe, for some good biological reason, might be relevant to the question you are asking. Trying every possible variable and hoping that some combination will produce a significant result is known as 'data dredging' and should be avoided at all costs. If you submit a paper for publication, you will be expected to justify your initial choice of variables.

Multivariate analysis of variance (MANOVA) extends ANOVA to include more than one dependent variable as well as several independent variables. So, for example, you might want to explore the relationships between newt numbers of two different species (the dependent variables) and a range of pond characteristics (size, depth, vegetation cover and so on) as independent variables. Do the two species have similar or different habitat requirements in the ponds? Do aspects of habitat interact in some way? MANOVAs can provide answers to such questions.

Multiple regression analysis allows the addition of many independent variables rather than the single one employed in linear regression to explain the distribution of a dependent variable. Does some combination of climatic factors influence the spawning date of frogs? Spawn date in this example is the dependent variable measured over many years in a particular pond. Climatic influences might include daytime temperature in the spawning month, rainfall that month or the month before, sunshine hours, and so on. A useful first step is first to check for pairwise correlations among the 'independent' variables that you think might be important, and if significant correlations are found, to discard one of the pair because it will essentially duplicate any explanation offered by the other. The remaining truly independent variables are then included in the analysis in various combinations, each of which generates an R^2 value with associated probability automatically adjusted for multiple variables. The objective is to determine which combination of independent variables best explains the distribution of the dependent variable. A regression equation identifying the relative strengths of these variables is part of the output. This kind of multivariate analysis is widely used in ecological research.

Another common objective is the identification of factors that determine whether a species is likely to be present or absent in a particular location. What combination of factors gives the highest probability that great crested newts will be found in a pond? In this case appropriate analytical tools are discriminant analysis and logistic regression, the second of which is increasingly popular. The dependent variable is categorical, one of just two states; in this example, species present or absent. The independent variables recorded from every pond where the newts are present and where they are absent are then sorted into two clusters to generate the best possible distinction between the two pond classes. These variables can be either 'continuous' (for example, the surface area of the pond), or categorical (for example, fish absent or present) or any combination of the two. Output includes not just the variable combinations but also the success rate of the best groupings, perhaps 'eighty percent of ponds with newts present and sixty percent of ponds with newts absent were correctly classified'. Ideally of course you hope for values approaching one hundred percent in both cases.

Although many people find statistics daunting they can actually be fun or at least very satisfying when results show significant effects and your hypothesis is supported. A friend and colleague, when asked by his research students why he never used his computer for games, responded that he got his kicks from MINITAB, a statistical software package. I know exactly what he means.

9.7 Publication outlets

If you have findings that are novel and based on observations that are carefully made and scrupulously recorded, your work may be worth publishing. There are several journals that publish original observations or research relating to amphibians and reptiles. In large part the choice of outlet depends on the type of work. Peer-reviewed journals require substantial scientific rigour in the design and execution of a submission. Every journal has its own format requirements and it is important to follow them exactly (instructions are provided on the websites, given below for the organisations publishing the most relevant journals). Writing scientific papers is an acquired skill and you should consider seeking help from a professional friend or colleague if you are completely new to it. Figures and photos in particular need to be very clear and software packages (such as Excel) should always be employed to generate graphs.

Journals suitable for relatively minor contributions include two published by the British Herpetological Society (BHS, see above) and one published by the Societas Europaea Herpetologica, SEH (European Herpetological Society).

These are:

The Natterjack newsletter, published a variable number of times every year by the BHS. Only members of the society have access to this.

The *Natural History Notes* section of *The Herpetological Bulletin*. Information for authors is provided on the Society's website. Free to BHS members but otherwise you will have to pay for a copy, although authors receive free pdf files of any paper they publish.

Herpetology Notes is published online by SEH. Details of submission requirements are available on the SEH website. Publication is free and pdfs are provided.

More substantial research, subject to peer review prior to acceptance, also has two outlets via the BHS and one other major option via SEH.

The main section of *The Herpetological Bulletin*, produced quarterly, publishes papers reporting original research but of limited extent.

The Herpetological Journal. Also published (quarterly) by the BHS as its mainstream high-quality science journal. Instructions for authors are provided on the BHS website. The Journal is free to members and pdfs of published papers are provided for authors.

Amphibia-Reptilia is published quarterly by SEH. It is free to members online and as hard copy; non-members are required to pay for access. Instructions for authors are on the SEH website.

For results directly applicable to conservation, especially where these relate to validation of methods, *Conservation Evidence* is a web-based journal well worth considering. It has a website (www.conservationevi-dence.com) and can be contacted for advice via info@conservationevidence.com.

This is far from an exhaustive list but it includes the most straightforward, close-to-home options. There are three mainstream North American herpetological journals which publish papers from around the world: *Copeia*, *Journal of Herpetology* and *Herpetologica*. All have websites with the required information for authors. Then there are more general journals that regularly include papers on amphibians and reptiles. *The Journal of Zoology*, published by the Institute of Zoology in London, is a good choice. Journals focusing on conservation research include *Animal Conservation*, *Biological Conservation* and *Conservation Biology*. However, these are increasingly hard to publish in and have high rejection rates because they are so popular and widely read. They may be worth a try if you are confident about a set of robust, novel results, but these journals also look for wide application, meaning that studies based just on a single species are unlikely to be accepted for publication unless you can demonstrate general implications (such as an impact of grazing across the board).

Finally, *British Wildlife* (www.britishwildlife.com) publishes authoritative, well written articles about all

British plants and animals and may be worth a look. Unlike the standard scientific outlets, this one actually pays authors for accepted work.

9.8 Meetings

It's always fun to get together with like-minded people and in the UK there are two annual meetings on amphibians and reptiles specifically designed to this end. Details of both are advertised on the ARC website months in advance. There is a registration fee, with coffee provided during intervals, and the website also gives information about local accommodation.

The Annual Joint Scientific Meeting is organised by ARC and BHS in early December, normally in Bournemouth. It comprises several talks over a single day, usually a Sunday. Contributions are based on scientific studies.

The Herpetofauna Recorders Meeting is organised by ARC, with help from ARG-UK, at the end of January or early in February. The venue moves around the UK. There are several talks on a Saturday, mostly about examples of conservation work, and workshops on various aspects of amphibian and reptile study on the Sunday (the second day is optional).

10 Useful addresses and links

10.1 Organisations

Contact details are shown below for a range of organisations that can prove information useful in various ways for anyone setting out to work with amphibians and reptiles. Bear in mind that these details (including, sometimes, even the organisation's name) may change over time but all should be recoverable by web searches.

Amphibian and reptile specialists

Amphibian and Reptile Conservation (ARC). 655A Christchurch Rd, Boscombe, Bournemouth, Dorset BH1 4AP. 01202 391319. enquiries@arc-trust.org. www.arc-trust. org. The largest and most active non-government organisation concerned with the conservation of amphibians and reptiles in Britain, including scientific research and data analysis. A repository for survey information including the NARRS scheme (www.arc-trust.org/science/narrs.php). Run by full-time professional staff. Has a 'friends' organisation but no formal membership.

 British Herpetological Society (BHS). 11 Strathmore Place, Montrose, Angus DD10 8LQ. info@thebhs.org. www.thebhs.org. Leading British membership organisation for all aspects of the study of amphibians and reptiles including their care in captivity. Run entirely by voluntary staff. Two major publications, both quarterly (Herpetological Bulletin and The Herpetological Journal).

 Amphibian and Reptile Groups of the UK (ARG UK). www.arguk.org. Parent organisation of all UK local (county-based) amphibian and reptile groups. Run entirely by voluntary staff. Use this link to find and contact your local group.

 Froglife. www.froglife.org. Non-government organisation promoting interest in amphibians and reptiles.

 Reptiles and Amphibians of the UK (RAUK). www.herpetofauna.co.uk. Web-based forum for exchange of observations and ideas on all aspects of amphibian and reptile biology, ecology and conservation. Run and updated on a voluntary basis.

 Societas Europaea Herpetologica (SEH). www.seh-herpetology.org. European herpetological membership society comparable to the British Herpetological Society, with a strong scientific remit. Organises scientific meetings at various venues around Europe every two years and publishes a scientific journal, *Amphibia-Reptilia*. Also has a conservation committee.

 Reptielen Amfibiien Vissen Onderzoek Nederland (RAVON). www.ravon.nl. A professional organisation dedicated to the conservation of amphibians, reptiles and fishes in the Netherlands. With a strong scientific background, RAVON is perhaps the most similar organisation in Europe to the UK's ARC (see above) and well worth a look to see how our colleagues abroad study and conserve amphibians and reptiles.

Other relevant non-government organisations (NGOs)

The Wildlife Trusts. www.wildlifetrusts.org. Parent organisation of the County Wildlife Trusts (CWTs) with more than forty local organisations spread across the UK. CWTs employ professional staff and often own and/or manage local nature reserves and other wildlife sites. Very useful sources of information about local amphibian and reptile distributions and recommended contacts, especially where a county has no ARG.

Pond Conservation. info@pondconservation.org.uk; 01865 483249. www.pondconservation.org.uk. A charity with full-time staff which, as its name suggests, actively promotes pond conservation (including creation and restoration) across the UK. Inevitably this includes work that benefits amphibians. This links to the European Pond Conservation Network http://campus.hesge.ch/epcn/contact.asp, which, also as its name suggests, is an organisation dedicated to looking after ponds at the pan-European scale.

Royal Society for the Protection of Birds (RSPB). www.rspb.org.uk. Britain's largest wildlife-dedicated non-government organisation with professional staff. The RSPB has nature reserves widely spread across Britain, many with substantial amphibian and reptile populations. These include the rarities (natterjack toads, sand lizards and smooth snakes) on coastal and heathland reserves.

National Trust (NT). www.nationaltrust.org.uk. Another very large non-government organisation with full-time staff. Although largely dedicated to preservation of Britain's architectural and cultural heritage, the NT also owns and manages areas of countryside to promote and sustain wildlife. As with the RSPB, many of these sites have amphibians and reptiles including, again, some with the rare species.

Wildfowl and Wetlands Trust (WWT). www.wwt.org.uk. An organisation primarily concerned with bird conservation but which has Wetland Centres around the UK in which proactive habitat conservation by professional staff also helps to sustain amphibian populations.

Statutory organisations
Government conservation agencies play vital roles in amphibian and reptile conservation. They sustain national nature reserves and extensive networks of protected sites, especially Sites of Special Scientific Interest (SSSIs) and provide funds for use by non-government organisations to assist with national conservation objectives. They include the licensing agencies for authorising work on highly protected species and on introduced species where release is required. Addresses of central offices (there are many regional ones) for licence applications are given below.

Central offices for licence applications
Countryside Council for Wales (CCW). www.ccw.gov.uk/?lang=en. Contact Species Protection Section, Countryside Council for Wales, Maes Y Ffynnon, Penrhosgamedd, Bangor, Gwynedd LL57 2DW. 0845 1306229.

Natural England (NE). Head Office is at Foundry House, 3 Millsands, Riverside Exchange, Sheffield S3 8NH. 0300 060 2745. www.naturalengland.org.uk. Licence application forms are available via the website. e-mail enquiries@naturalengland.org.uk

Northern Ireland Environment Agency (NIEA). 0845 3020008. Contact via www.doeni.gov.uk/niea/biodiversity/sap_uk/wildlife.htm

Scottish Natural Heritage (SNH). Head Office is at Great Glen House, Leachkin Road, Inverness IV3 8NW. 01463 725000. www.snh.gov.uk Licence application forms are available via the website.

The Environment Agency (EA). National Customer Contact Centre, PO Box 544, Rotherham S60 1BY. 03708 506506. www.environment-agency.gov.uk. E-mail enquiries@environment-agency.gov.uk. A useful organisation to consult about research or fieldwork likely to involve freshwater fish.

Equipment suppliers

There are many retail outlets for the kind of equipment useful for amphibian and reptile research. Below are just a few examples.

Aquaria and vivaria

ARKpetsonline. www.arkpetsonline.co.uk sells a wide range of vivaria and plastic tanks suitable for experiments with amphibians and reptiles.

Pond and aquarium nets

NHBS - Everything for widlife, science & environment. www.nhbs.com.
Efe & GB Nets. www.efe-gbnets.com

Powerful torches

Cluson Engineering. Unit 6, Bedford Rd, Petersfield, Hants GU32 3LJ. 01730 264672. http://clulite.cluson.co.uk. e-mail sales@clulite.co.uk. Excellent (but not cheap) powerful torches with rechargeable batteries, especially the Clubman range (1 million candlepower).

Food for captive amphibians and reptiles

Daphnia can be caught in local ponds and are often available in local aquarist shops around the country. So are other invertebrate foods such as mealworms and crickets. There are however some useful big suppliers of crickets, mealworms and frozen mice. A good example is Live Foods Direct (www.livefoodsdirect.co.uk) which sells a comprehensive range.

References

Arnold, N. & Ovenden, D. (2002) *A field guide to reptiles and amphibians of Britain and Europe* (2nd edition). HarperCollins, London.

Baker, J.M.R. (1999) Abundance and survival of great crested newts (*Triturus cristatus*) at a pond in central England: monitoring individuals. *Herpetological Journal* 9, 1–8.

Baker, J., Beebee, T., Buckley, J., Gent, T. & Orchard, D. (2011) *Amphibian habitat management handbook*. Amphibian and Reptile Conservation, Bournemouth.

Bardsley, L. & Beebee, T.J.C. (1998) Interspecific competition between *Bufo* larvae under conditions of community transition. *Ecology* 79, 1751–1760.

Beebee, T.J.C. (1979) Habitats of the British amphibians (2): suburban parks and gardens. *Biological Conservation* 15, 241–257.

Beebee, T.J.C. (1996) *Ecology and conservation of amphibians*. Chapman & Hall, London.

Beebee, T.J.C. (2005) Amphibian conservation genetics. *Heredity* 95, 423–427.

Beebee, T.J.C. & Griffiths, R.A. (2000) *Amphibians and reptiles*. HarperCollins, London.

Beebee, T.J.C., Wilkinson, J.W. & Buckley, J. (2009) Amphibian declines are not uniquely high amongst the vertebrates: trend determination and the British perspective. *Diversity* 1, 67–88.

Cooke, A.S. (2011) The role of road traffic in the near extinction of common toads in Ramsey and Bury. *Nature in Cambridgeshire* 53, 45–50.

Cooper, W.E., Perez-Mellado, V. & Vitt, L.J. (2004) Ease and effectiveness of costly autotomy vary with predation intensity among lizard populations. *Journal of Zoology* 262, 243–255.

Davies, N.B. & Halliday, T.R. (1977) Optimal mate selection in the toad *Bufo bufo*. *Nature* 269, 56–58.

Denton, J.S. & Beebee, T.J.C. (1993) Density-related features of natterjack toad (*Bufo calamita*) populations in Britain. *Journal of Zoology* 229, 105–119.

Duellman, W. E. & Trueb, L. (1994) *Biology of amphibians* 2nd edition. The John Hopkins University Press, Baltimore.

Dytham, C. (2010) *Choosing and using statistics: a biologist's guide* (3rd edition). Blackwell, Oxford.

Edgar, P., Foster, J. & Baker, J. (2010) *Reptile habitat management handbook*. Amphibian and Reptile Conservation, Bournemouth.

Forsman, A. & Lindell, L.E. (1997) Responses of a predator to variation in prey abundance: survival and emigration of adders in relation to vole density. *Canadian Journal of Zoology* 75, 1099–1108.

Gabor, C.R. & Halliday, T.R. (1997) Sequential mate choice by multiply mating smooth newts: females become more choosy. *Behavioural Ecology* 8, 162–166.

Gosner, K.L (1960) A simplified table for staging anuran embryos and larvae with notes on identification. *Herpetologica* 16, 183–190.

Griffiths, R.A. (1985) A simple funnel trap for studying newt populations and an evaluation of trap behaviour in smooth and palmate newts *Triturus vulgaris* and *T. helveticus*. *Herpetological Journal* 1, 5–9.

Griffiths, R.A. (1986) Feeding niche overlap and food selection in smooth and palmate newts, *Triturus vulgaris* and *T. helveticus*, at a pond in mid-Wales. *Journal of Animal Ecology* 55, 201–214.

Griffiths, R.A. (1996) *Newts and salamanders of Europe*. Poyser Ltd, London.

Griffiths, R.A., Sewell, D. & McCrea, R.S. (2010) Dynamics of a declining amphibian metapopulation: survival, dispersal and the impact of climate. *Biological Conservation* 143, 485–491.

Hedges, S.B. & Poling, L.L. (1999) A molecular phylogeny of reptiles. *Science* 283, 998–1001.

Inns, H. (2009) *Britain's reptiles and amphibians*. Wild Guides, Old Basing, Hampshire.

Jehle, R., Thiesmeier, B. & Foster, J. (2011) *The crested newt*. Laurenti-Verlag, Bielefeld.

Langton, T.E.S. (1989) *Amphibians and roads*. ACO Polymer Products, Shefford.

Madsen, T., Stille, B. & Shine, R. (1996) Inbreeding depression in an isolated population of adders *Vipera berus*. *Biological Conservation* 75, 113–118.

Maitland, P.S. & Campbell, R.N. (1992) *Freshwater Fishes*. HarperCollins, London.

Malmgren, J.C. (2002) How does a newt find its way from a pond? Migration patterns after breeding and metamorphosis in great crested newts (*Triturus cristatus*) and smooth newts (*T. helveticus*). *Herpetological Journal* **12**, 29–35.

McDiamid, R.W. & Altig, R. (1999) *Tadpoles*. University of Chicago Press, Chicago.

McGarigal, K., Cushman, S. & Stafford, S. (2002) *Multivariate statistics for wildlife and ecology research*. Springer, London.

Meyer, A.H., Schmidt, B.R. & Grossenbacher, K. (1998) Analysis of three amphibian populations with quarter-century long time-series. *Proceedings of the Royal Society B* **265**, 523–528.

Mole, S.R.C. (2010) Changes in relative abundance of the western green lizard *Lacerta bilineata* and the common wall lizard *Podarcis muralis* introduced onto Boscombe Cliffs, Dorset. *Herpetological Bulletin* **114**, 24–29.

Mullin, S.J. & Seigel, R.A. (2009) *Snakes: ecology and conservation*. Comstock Publishing Associates, Cornell University Press.

Oldham, R.S., Keeble, J, Swan, M.J. & Jeffcote, M. (2000) Evaluating the suitability of habitat for the great crested newt (*Triturus cristatus*). *Herpetological Journal* **10**, 143–155.

Pedersen, I.L., Jensen, J.K. & Toft, S. (2009) A method of obtaining dietary data for slow worms (*Anguis fragilis*) by means of non-harmful cooling and results from a Danish population. *Journal of Natural History* **43**, 1011–1025.

Pellet, J. & Schmidt, B.R. (2005) Monitoring distributions using call surveys: estimating site occupancy, detection probabilities and inferring absence. *Biological Conservation* **123**, 27–35.

Platenberg, R.J. & Griffiths, R.A. (1999) Translocation of slow-worms (*Anguis fragilis*) as a mitigation strategy: a case study from south-east England. *Biological Conservation* **90**, 125–132.

Pyron, R.A. & Wiens, J.J. (2011) A large-scale phylogeny of Amphibia including over 2800 species, and a revised classification of extant frogs, salamanders, and caecilians. *Molecular Phylogenetics & Evolution* **61**, 543–583.

Reading, C.J. (2004) Age, growth and sex determination in a population of smooth snakes, *Coronella austriaca*, in southern England. *Amphibia-Reptilia* **25**, 137–150.

Reading, C.J. (1997) A proposed standard method for surveying reptiles on dry lowland heath. *Journal of Applied Ecology* **34**, 1057–1069.

Reading, C.J. (2010) Is cattle grazing harmful to smooth snakes on lowland dry heath? *Herpetological Journal* **20**, 3.

Reading, C.J. & Jofre, G.M. (2009) Habitat selection and range size of grass snakes *Natrix natrix* in an agricultural landscape in southern England. *Amphibia-Reptilia* **30**, 379–388.

Reilly, S.M., McBrayer, L.B. & Miles, D.B. (eds) (2007). *Lizard ecology*. Cambridge University Press, Cambridge.

Roberts, J.M. & Griffiths, R.A. (1992) The dorsal stripe in newt efts: a method for distinguishing *Triturus vulgaris* and *T. helveticus*. *Amphibia-Reptilia* **13**, 13–19.

Rowe, G. & Beebee, T.J.C. (2007) Defining population boundaries: use of three Bayesian approaches with microsatellite data from British natterjack toads (*Bufo calamita*). *Molecular Ecology* **16**, 785–796.

Sewell, D., Beebee, T.J.C. & Griffiths, R.A. (2010) Optimising biodiversity assessments by volunteers: the application of occupancy modelling to large-scale amphibian surveys. *Biological Conservation* **143**, 2102–2110.

Stuart, S.N., Chanson, J.S., Cox, N.A., Young, B.E., Rodrigues, A.S.L., Fischmann, D.L. & Waller, R.W. (2004) Status and trends of amphibian declines and extinctions worldwide. *Science* **306**, 1783–1786.

Teacher, A.G.F., Cunningham, A.A. & Garner, T.W.J. (2010) Assessing the long-term impact of Ranavirus infection in wild common frog populations. *Animal Conservation* **13**, 514–522.

Vitt, L.J. & Caldwell, J.P. (2009) *Herpetology: an introductory biology of amphibians and reptiles*. 3rd edition. Elsevier, Boston.

Wells, K.D. (2007) *The ecology and behaviour of amphibians*. Chicago University Press, Chicago.

Wheater, C.P. & Cook, P.A. (2003) *Studying invertebrates*. Naturalists' Handbook no. 28. The Richmond Publishing Co. Ltd, Slough.

Wycherley, J. & Anstis, R. (2001) *Amphibians and reptiles of Surrey*. Surrey Wildlife Trust, Woking.

Zuiderwijk, A. & Janssen, I. (2008) Results of 14 years reptile monitoring in the Netherlands. *Proceedings of the Sixth World Congress of Herpetology, Manaus, Brazil*. Available from RAVON, http://www.ravon.nl

Index

Bold page numbers indicate the colour plates in chapter 8. Page numbers in *italics* refer to tables or figures.